A. E. Kennelly, Edwin J. Houston

Electricity in Electro-Therapeutics

A. E. Kennelly, Edwin J. Houston
Electricity in Electro-Therapeutics
ISBN/EAN: 9783337406165

Printed in Europe, USA, Canada, Australia, Japan

Cover: Foto ©berggeist007 / pixelio.de

More available books at **www.hansebooks.com**

BY THE SAME AUTHORS

Elementary Electro-Technical Series

COMPRISING

Alternating Electric Currents.
Electric Heating.
Electromagnetism.
Electricity in Electro-Therapeutics.
Electric Arc Lighting.
Electric Incandescent Lighting.
Electric Motors.
Electric Street Railways.
Electric Telephony.
Electric Telegraphy.

Cloth,   Price per Volume, $1.00.

Electro-Dynamic Machinery.
Cloth, $2.50.

THE W. J. JOHNSTON COMPANY
253 Broadway, New York

ELEMENTARY ELECTRO-TECHNICAL SERIES

ELECTRICITY
IN
ELECTRO-THERAPEUTICS

BY

EDWIN J. HOUSTON, Ph. D.

AND

A. E. KENNELLY, Sc. D.

NEW YORK

THE W. J. JOHNSTON COMPANY

253 BROADWAY

1896

COPYRIGHT, 1896, BY
THE W. J. JOHNSTON COMPANY.

CONTENTS.

CHAPTER		PAGE
I.	Introductory,	1
II.	Electromotive Force,	13
III.	Electric Resistance,	63
IV.	Electric Current,	80
V.	Varieties of Electromotive Force,	107
VI.	Electric Work and Activity,	124
VII.	Frictional and Influence Machines,	138
VIII.	Magnetism,	184
IX.	Induction of E. M. F. by Magnetic Flux,	221
X.	The Medical Induction Coil,	248

CHAPTER		PAGE
XI.	Dynamos, Motors and Transformers,	299
XII.	High Frequency Discharges,	329
XIII.	Electrolysis and Cataphoresis,	356
XIV.	Dangers in the Therapeutic Use of Electricity,	365
Index,		373

PREFACE

This little book, entitled Electricity in Electro-Therapeutics, is intended to meet a growing demand which exists not only on the part of general medical practitioners, but also on that of the general public, for reliable information respecting such matters in the physics of electricity applied to Electro-Therapeutics, as can be readily understood by those not specially trained in electro-technics.

Electricity has recently made such rapid strides in application to both surgical and medical practice, that recent information, concerning electrical developments in apparatus and in theory, is much in request by those interested in the healing art.

The method of treatment adopted throughout the book in the description of electro-technics has been the circuital method; that is to say, all the phenomena of electricity and magnetism have been considered as pertaining either to the electro-static, the electric, or the magnetic circuit, and the laws of these three circuits have been developed upon analogous lines. The authors believe that this treatment is the key-note to a clear comprehension of the numerous and often complex electro-magnetic phenomena met with in the application of electricity to electro-therapeutics.

In thus aiding the general public to readily comprehend the principles underlying the physics of electro-therapeutics, the authors trust that they are aiding the general cause of humanity in enabling electricity to be employed more intelligently in the healing art, as well as permitting truth to be discerned from fraud.

The authors therefore present this book to the public in the hope that it may prove serviceable.

ELECTRICITY IN ELECTRO-THERAPEUTICS.

CHAPTER I.

INTRODUCTORY.

It has not infrequently happened in the history of scientific discovery, that from various causes, the discoverer has obtained so incomplete a view of the new fact, as to entirely lose sight of its true significance, and to regard it merely from the standpoint of some of its unimportant characteristics. This was the case, in the celebrated discovery made by Luigi Galvani, in 1786, concerning the existence of what he at first believed to be the vital fluid, or essence of animal vitality, but which was afterwards proved by Volta to be essentially a new method of producing electricity by chemical action.

Galvani discovered that if the hind legs of a recently killed frog be deprived of

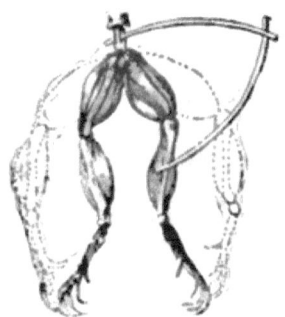

Fig. 1.—Galvanoscopic Frog.

their integument, and the lumbar nerves, suitably exposed in their position on either side of the vertebral column, be connected with the crural muscles by a metallic strip, as shown in Fig. 1, that these muscles will be brought into a spasmodic activity

closely resembling their action during life.
It has been alleged that this classic experiment of Galvani was the result of chance; that he had prepared some frogs' legs for supper, and, happening to hang them against an iron balcony, he noticed that the muscles twitched as soon as they touched the iron; that is, went convulsively through their motions as in life, and that these motions were repeated whenever the contact was renewed. Moreover, the power of producing these convulsive movements was retained by the limbs for an hour or more after removal from the body. This account would seem improbable, since Galvani was well aware of the fact that an electric discharge, sent through the legs of a recently killed frog, would produce convulsive movements in them; and, indeed, he was in the habit of employing such legs as a form of sensitive

galvanoscope, though, of course, Galvani
did not know it under this name, but he
did know that it formed a much more
sensitive apparatus for detecting an electric
current than the pith-ball electroscope employed in those early days, as almost the
only available means for detecting an
electric charge.

Unfortunately, Galvani failed to recognize the extreme importance of his discovery. Working, as he had been for a
long time, with the hope of discovering the
seat of animal vitality, he was only too
willing to find in this observation the
principle of that vital fluid for which he
so long and ardently had sought. He was,
therefore, handicapped in the search, and
unfitted, to a certain extent, to weigh
calmly the evidence presented. Though
thoroughly familiar with the convulsive

twitchings produced by the passage of the electric discharge through the frog's legs, it never seemed to occur to him that what he had in reality discovered, was an entirely new method of producing electric discharges. He only saw in this observation what he so ardently desired to see; namely, convulsive muscular movements, due to a vital fluid, which, he believed, came from the nerve of the animal, and was conveyed through the metallic conductor to the muscles, where it produced the characteristic twitchings.

The announcement by Galvani of his discovery, produced the most intense excitement throughout the scientific world, and his views as to the cause of the phenomenon were at first generally accepted. Among, perhaps, the most ardent of his early followers was

Alessandro Volta, who at once repeated Galvani's experiments, and began a series of extended researches on the phenomena. Volta soon reached the conclusion that the twitchings of the legs of Galvani's frogs were to be ascribed, not to a vital fluid, but to the presence of an electric discharge, and that, consequently, sight had been lost of the most important part of Galvani's experiment; namely, that it furnished a new method of producing electricity.

Volta showed, among other things, that the convulsive movements were more pronounced when the nerves were connected with the muscles by two dissimilar metals, instead of by a single metal, and ascribed the cause of the electricity produced as the *contact of dissimilar substances*. In the light of more modern scientific discovery, it would appear that both discoverers saw but a partial truth, although Volta had undoubtedly a more complete grasp of the phenomena than Galvani. Galvani observed the convulsive movements, but improperly attributed them to the presence of a vital fluid. Volta correctly ascribed the cause of the movements to the passage of electricity, but incorrectly ascribed the cause of the continuous supply of electric current to the contact of dissimilar substances. Modern research has shown that contact alone is unable to account for the continuous production of an electric current, and that such a discharge occurs only under circumstances when chemical actions take place.

In endeavoring, at the present time, to gauge the value to the world of the discoveries of these two pioneer investigators, a tendency may exist to give, too unreservedly, the award to Volta, on the plea that the result of his investigations produced, some ten years later, the great discovery of the Voltaic pile, a discovery which has done so much for the world's progress, but it must not be forgotten that it was the original observations of Galvani, aided, it is true, largely by his afterwork in the same field, that first called attention to the wonderful effects which electricity produces on the animal organism, and if to-day electro-therapeutics, or the application of electricity for the restoration of the healthy condition of the body, is an actual power, the beneficial effects of which are apparently being more and more clearly established every year, it is undoutedly to Galvani that the guerdon of the discovery must be awarded.

Without attempting to trace the history of the extended experiments that were made on Galvani's original observation, experiments that have continued up to the present day, and without stopping to consider the extended and bitter controversy that was waged between the disciples of Galvani on one side, and those of Volta on the other, as to the cause of the phenomena, or the equally extended controversy which existed as to the origin of the electric current produced in the voltaic pile, or cell, it will suffice, for our present purpose, to consider animals as electric sources, and the effects produced by the passage of electricity through animals. These can be briefly summarized as follows; viz.,

(1) That the body of an animal is, in itself, the seat of electric currents.

(2) That these currents exist not only during the abnormal or diseased condition

of different parts of the body, but also in the normal condition of the body.

(3) That electric discharges, when sent through the body of an animal, are capable of producing marked effects therein, the character of which depends upon the nature of the discharge.

(4) That the passage of an electric discharge through a nerve, muscle, or indeed through any organ of the body of an animal, produces an alteration in its functional activity.

(5) That a sufficiently powerful discharge through the body of an animal may produce death.

The effects of electricity on the human body have been very generally recognized since the time of Galvani, and it is generally believed that electricity possesses powerful remedial properties; but, like

ELECTRO-THERAPEUTICS. 11

all remedial agencies, unless intelligently applied, it may produce more harm than good. That much harm has been done by its improper use there can be no doubt. Indeed, one of the results of such misuse has been even at the present time to greatly retard its general introduction. Much of this difficulty has arisen from want of a general knowledge of the fundamental laws underlying the production and action of electricity and magnetism. As a result of a lack of such knowledge, extravagant and ridiculous claims are frequently made as to the wonderful curative powers of certain apparatus, alleged to produce electric currents; apparatus that even a tyro in electricity would at once be able to show are incapable of producing any current whatever.

A knowledge of the laws of electricity

and magnetism will, therefore, not only enable the general public to detect electro-therapeutic frauds, but will also tend to increase confidence in legitimate electro-therapeutic applications, a confidence justly merited by **experience.**

Since both electricity and magnetism are exact sciences, their application to the art of electro-therapeutics cannot fail to be of benefit in scientific treatment.

CHAPTER II.

ELECTROMOTIVE FORCE.

ALTHOUGH we are ignorant of the exact nature of electricity, yet it is by no means true that we are ignorant of the laws under which electricity operates. In other words, our ignorance relates to the exact nature of the electric force rather than to its laws. In the physical world we can properly claim knowledge concerning a force, when we are able to predict its action under given conditions. Gauged in this way, our knowledge of electricity is both extensive and accurate, since it is possible to fairly predict just what will happen under a great variety of electric conditions.

It will be generally acknowledged, however, that we certainly know this much about electricity; viz., that it cannot be regarded as a form of matter; at least not matter in the ordinary sense of the word. All matter with which we are acquainted, exercises gravitational influence, that is to say, tends to attract other matter towards it. No such tendency has yet been shown to exist in the case of electricity. But ordinary matter is not the only material with which we are surrounded. The entire universe is believed to be pervaded by a highly tenuous medium called the *ether*, which transmits light, heat and gravitation.

As frequent reference will be made in this book to the existence and properties of the ether, it may be well to explain the nature of the evidence which has convinced scientific men of its existence. We know that the effects produced by a sounding body, such, for example, as a vibrating bell, are transmitted across the space between the bell and the observer's ear, by means of waves, or to-and-fro

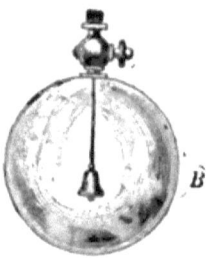

Fig. 2.—Transmission of Sound through a Vacuum.

motions produced in the medium existing between the bell and the ear. This medium is usually the air. If a bell, placed inside a glass vessel, as shown in Fig. 2, be set vibrating, it can be heard by an observer at some distance, since the air the vessel contains transmits the vibrations to the sides of the globe, which, in their turn, transmit the vibrations to the external air, and so to the observer's ear. But **if the vessel** be exhausted; *i. e.*, **deprived of its air,** the **sound** will no longer be transmitted, since the medium is then removed which carries its vibrations. If, however, a similar **experiment be tried with a hot body, it will be** found **that the** existence **of such a** medium as air is **not essential to** permit the body to transmit **its heat across** intervening space. For example, **if, as in** Fig. 3, two reflectors *A* and *B*, **be placed** inside the glass receiver of an air-pump, and a delicate thermometer *T*, be suitably supported **at the focus of one of these** reflectors, and a platinum wire be placed **at the focus of the other** reflector, then, **if** an electric current be sent **through the** platinum wire, of such strength as to **render** it incandescent, the **heat** radiated **from** the wire will **be** reflected successively from

A and B, and be focused on the thermometer, which will immediately indicate an increased temperature, and this effect will occur, whether the receiver contains

Fig. 3. Transmission of Heat through a Vacuum.

air or is exhausted. Evidently, the presence of a gross medium like air is not essential to the transmission of radiant heat, and in this respect differs from the transmission of sound just referred to. Moreover, the light emitted by the glowing platinum wire is also transmitted through the empty space in the globe and renders the wire visible.

A little reflection will show that the preceding experiment is not, in reality, needed to prove the possibility of radiant light and heat being readily transmitted across space devoid of ordinary matter; for, radiant light and heat reach the earth from the sun, and from the fixed stars, across space existing between the earth and the heavenly bodies, which space we believe to be devoid of ordinary matter. In the early history of physical science this fact led to the belief that light and heat were effects produced by specific fluids or effluvia; that a hot body sent off a specific effluvium which constituted heat; and, in a similar manner, a luminous body sent off a specific effluvium constituting light. The particular effluvium in the case of heat received the name of *caloric*, a term which, unfortunately, even to-day, is still loosely employed in the science of heat. Without entering into minute details, it suffices to say that the theory which ascribes light or heat to the existence of special fluids is now considered absolutely untenable. Light and heat, like sound, are believed to be produced by vibrations, or to-and-fro motions. A medium, therefore, is necessary to carry the vibrations of light and heat, and, consequently, the highest vacuum which can be obtained is, for this reason, believed to be filled with ether, that is to say, the ether is not pumped out of a reservoir by the action of the air-pump.

The ether is sometimes called the *luminiferous ether* because it transmits the vibrations of light. It is also called the *universal ether* because it is supposed to exist everywhere. Even in the densest of solid bodies it is assumed to exist, between the ultimate particles, of which such bodies are formed; namely, between the atoms and the molecules.

Electricity and magnetism, like heat and light, are also capable of manifesting their influence through the air-pump vacuum. For example, the filament of an incandescent lamp, which, as is well known, is placed in a very high vacuum, can, nevertheless, be deflected by the electric attraction produced by a charged body *A*, Fig. 4, at some distance from the lamp. Here the electric attraction traverses the apparently empty space surrounding the filament, and, consequently, must have acted through the ether which

Fig. 4.—Transmission of Electrostatic Force Across a Vacuum.

is believed to fill the exhausted lamp chamber.

In a similar manner, magnetic attraction is capable of acting through empty space;

for, if, as in Fig. 5, an incandescent electric lamp be brought near a powerful magnet, when a continuous current is passing

Fig. 5.—Transmission of Magnetic Force Across a Vacuum.

through the filament, the filament has thereby acquired magnetic properties, and a deflection of the filament will take place, although no medium, save the ether, can apparently convey this influence within the chamber. Here, as before, this particular experiment is not necessary to show that magnetic influence can traverse apparently empty space; for, during the prevalence of an unusual number of spots on the surface of the sun, there are produced marked disturbances on delicately suspended compass needles on the earth. This influence is apparently transmitted through the ether which we believe fills interstellar space.

It is interesting to note in this connection, that the early views concerning the nature of electricity regarded it as a fluid or fluids, just as heat and light were originally regarded. It is now the belief, however, that electricity and magnetism are phenomena connected with some active condition of the ether. For example, light is almost universally regarded as being transmitted by a particular transverse vibration of the ether, and some particular forms of disturbances in the ether are also believed to be the causes of electric and magnetic phenomena.

Before, however, discussing at greater length the nature of electricity, let us consider a well-known electric source, as, for example, the dynamo-electric machine, such as is used for generating electric currents for arc or incandescent lights. Here, popularly, the machine is spoken of as producing electricity. Strictly speaking, however, it is not electricity which the machine primarily produces. The machine produces a variety of force capable of starting, or producing, an electric current under suitable conditions. This force produced by the dynamo is termed *electromotive force*, or the force which tends to set electricity in motion, so as to cause an electric flow. It is essential to bear in mind that in no case can electricity be produced by any machine without the prior production of electromotive force, just as no motion can exist in any material object without the antecedent application of a material force; *i. e.*, of a body-moving force. Whenever, therefore, electricity is produced, no matter what the nature of the machine, or source producing it may be, the machine or source must necessarily first produce an electromotive force, and this electromotive force in its turn will or will not produce electricity, according to the conditions under which it acts. Electromotive force is usually abbreviated E. M. F.

Various devices are employed in practice for the production of E. M. F. Such devices may be classified as follows:

(1) Those produced by chemical action; such as a voltaic cell or primary cell, and a charged storage cell or secondary cell.

(2) Those produced by the action of radiant energy; *i. e.*, radiant light or heat; such, for example, as a thermo-electric cell.

(3) Those produced by the action of mechanical energy; such, for example, as a dynamo-electric machine, a frictional machine, an electrostatic induction machine, or a liquid flowing through a capillary tube.

(4) Those produced by vital energy, such as an animal or a plant regarded as an electric source.

As already stated, in any of the preceding sources it is E. M. F. which is primarily produced.

ELECTRO-THERAPEUTICS. 27

Take, for example, a form of voltaic cell, shown in Fig. 6, known as the *Leclanché*

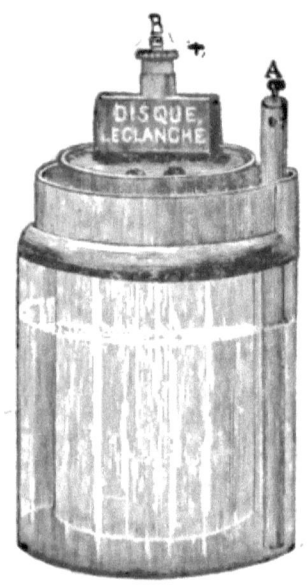

FIG. 6.—LECLANCHÉ VOLTAIC CELL.

cell. Here it is the chemical energy of combination between the zinc plate A, and the solution of ammonium chloride, which enables the electric current to be sustained, but the cell always produces an E. M. F., although it will not supply an electric current until its terminals *A* and *B*, are connected together by means of a conductor or external circuit.

In order to measure the E. M. F. of any electric source, a *unit of E. M. F.* has been internationally adopted. This unit is called the *volt*, after Alessandro Volta, the inventor of the voltaic cell. The E. M. F. of the Leclanché cell shown in Fig. 6, is, approximately, 1½ volts, and that of the ordinary blue-stone cell, as shown in Fig. 7, is about one volt.

In both the Leclanché and the blue-stone voltaic cells, it will be observed that there are two metallic substances immersed in a liquid. For example, in the Leclanché cell, shown in Fig. 6, the two substances are carbon and zinc, and the solution in which they are plunged is an aqueous solution of sal-ammoniac. In the *bluestone*, or *gravity cell*, shown in Fig. 7, the

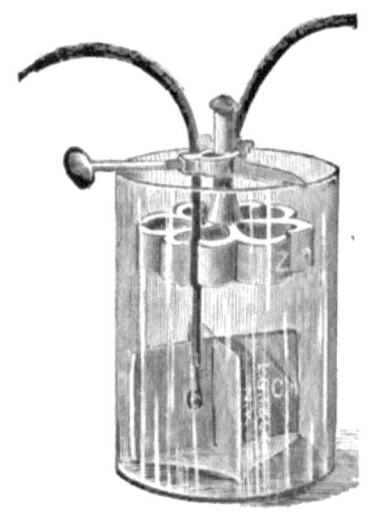

Fig. 7.—Bluestone or Gravity Voltaic Cell.

two metals are zinc and copper, marked Zn and Cu, but here there are two separate exciting liquids; namely, a dense solution of copper sulphate, which occupies the lower part of the cell, and a lighter solution of zinc sulphate, which surrounds the zinc plate and floats upon the copper sulphate solution. All voltaic cells may be divided into two general classes; namely,

(1) The *single-fluid* cells, or those which have a single exciting fluid; and,

(2) The *double-fluid* cells, or those which, like the bluestone cell, have two exciting fluids.

Every voltaic cell, whether of the double- or single-fluid type, consists of two essential parts; namely,

(1) Of a *voltaic pair* or *voltaic couple*, consisting of two dissimilar electrically conducting substances.

(2) Of an *exciting liquid* called the *electrolyte*, capable of conducting electricity, and of being decomposed by it. The double-fluid cells have two liquids or electrolytes. The two substances forming a voltaic pair or couple, are called the

elements of the cell. Voltaic elements are generally made in the form of plates or rods, and are known respectively as the *positive* and the *negative plates* or elements.

During the action of a voltaic cell, that is, while it is furnishing electric current to the circuit connected with it, a chemical action takes place between one or both of the electrolytes and one of the plates. The result of this action is that one of the plates, the positive, is gradually dissolved, or enters into chemical combination with part of the electrolyte, the other plate remaining unacted on. In nearly all forms of voltaic cells, there results from this decomposition a tendency to liberate hydrogen at the surface of the negative plate, or the plate which is unacted on. If hydrogen be permitted to be liberated on the surface of the negative plate, a marked decrease occurs in the ability of the cell to furnish current, for reasons which will be pointed out hereafter.

In single-fluid cells no provision is made to prevent the evolution of hydrogen at the negative plate; **or, as it** is generally **called,** the *polarization* of the negative plate. **In the** double-fluid **cell the second** fluid consists of a substance which surrounds the negative plate, **and is** provided for the express purpose of entering into combination with the hydrogen and so preventing its being liberated. There are some forms of voltaic cells which are apparently single-fluid cells, since they possess but a single fluid, or electrolyte, but which properly come under the type of double-fluid cells, since they are non-polarizable, **being** provided with a solid substance in contact with the negative plate, that is capable of combining with hydrogen and thereby preventing its liberation. For example, in the Leclanché cell, shown in Fig. 6, the elements of the voltaic couple are zinc and carbon, immersed in a solution of sal-ammoniac in water, but the carbon is surrounded by granulated, solid peroxide of manganese, which possesses the power of readily entering into combination with hydrogen.

A great variety of conducting substances are employed in pairs for the couples of voltaic cells. The most important of these, however, are zinc, carbon, copper, lead, silver, and platinum. Of these substances, zinc in nearly all cases forms the positive element; that is to say, it forms a voltaic couple either with carbon, copper, lead, silver, or platinum.

When a voltaic cell has its *circuit closed*, for example, when the zinc-carbon couple shown in Fig. 6, is connected to an external circuit, the E. M. F. it produces causes a current to flow through such circuit. For purposes of convenience it has been agreed, conventionally, to regard the electric current as leaving a voltaic cell at a particular point, and, after passing through the circuit, to re-enter it at another point. The point at which the current is conventionally assumed to leave the cell is called its *positive pole*, and the point at which it is assumed to re-enter it, after having passed through the circuit, the *negative pole*. The positive pole of the cell is the pole connected with the plate which is not acted on, that is with the negative plate, while the negative pole is the pole connected with the plate which is chemically acted on, or the positive plate. For example, in the battery shown in Fig. 7, the positive pole is connected with the copper plate, and is marked with a plus, and an arrow, indicating the fact that the current leaves the cell at this pole; while the negative pole is the terminal of the zinc plate, and is indicated by a minus sign, and an arrow flowing towards the cell, indicating the fact that the current enters the cell at this pole. In the battery shown in Fig. 6, the positive pole is the terminal B, of the carbon element, while the negative pole is the terminal A, of the zinc element.

Voltaic cells form an important electric source much employed in electro-therapeutics. We will, therefore, briefly describe some of their more important practical forms.

Fig. 8, shows a form of voltaic cell,

called the silver-chloride cell. This cell
consists of a zinc-silver couple, immersed

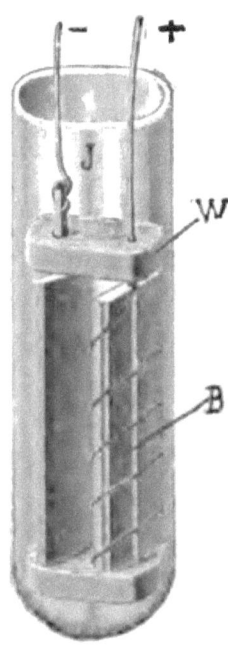

Fig. 8.—Form of Silver-Chloride Cell.

in a dilute aqueous solution of sal-ammoniac. The silver plate has the form of
a wire, and is surrounded by a fused mass
of silver chloride. The arrangement of

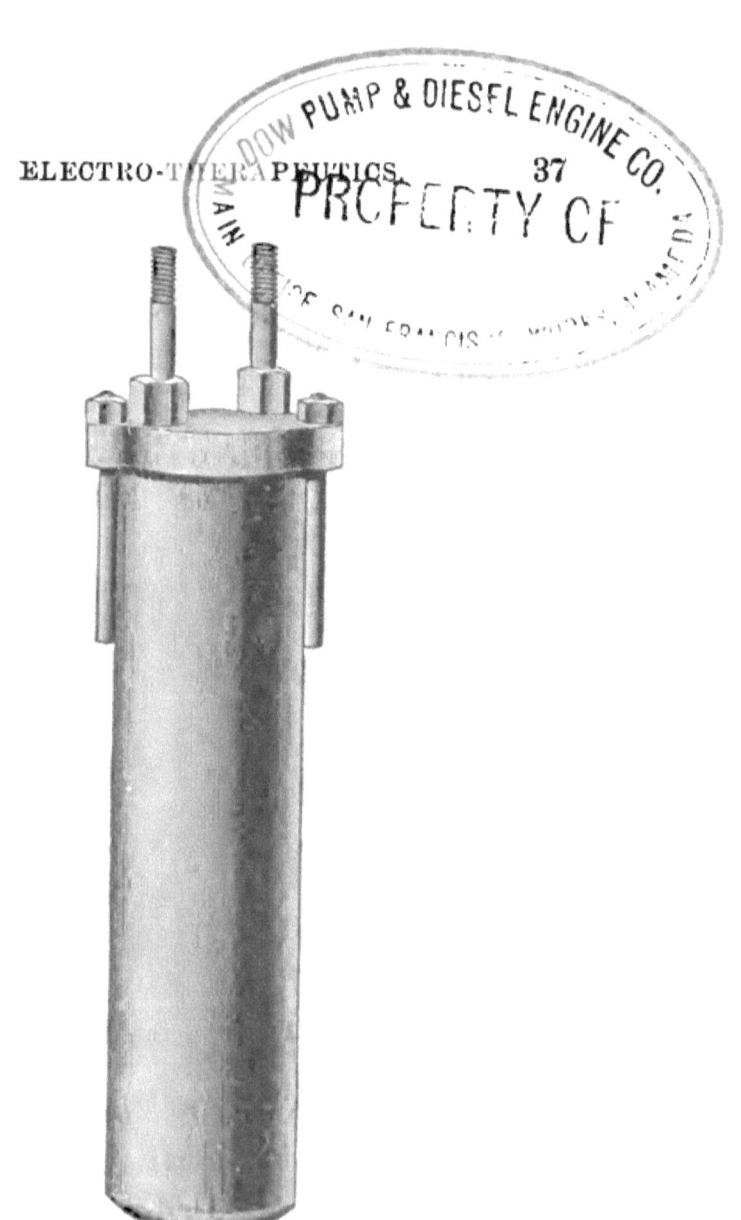

Fig. 9.—Silver-Chloride Voltaic Cell.

the plates is shown in the figure. It will be seen that a thread B, is wrapped around the silver and silver chloride, so as to prevent the possibility of contact between the silver element and the zinc plate. The two plates are also kept apart by a small block of wood W. The couple so formed is placed in a small glass or rubber jar J, containing the exciting solution of sal-ammoniac. Another form of silver-chloride cell is shown in Fig. 9. The advantage of the silver-chloride cell consists in its portability. As many as fifty of these cells can be set in a frame, as shown in Fig. 10, and enclosed in a small wooden box weighing only ten pounds. The silver-chloride cell is very nearly uniform in its electromotive force, which has a value of about 1.03 volts. The cell possesses, however, the disadvantage of not being able to supply powerful currents continuously, owing to the fact that in order to obtain portability, its elements are made so small. Were the cell constructed of such a size as would permit it to supply

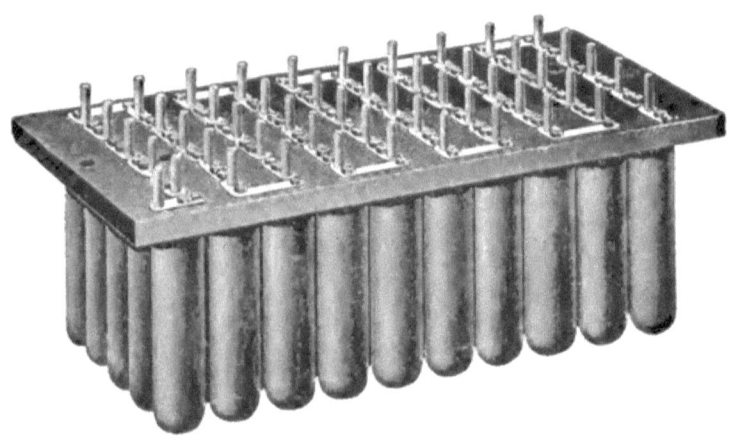

FIG. 10.—BATTERY OF SILVER-CHLORIDE CELLS.

powerful currents, the cost of the silver and silver-chloride would render its use impracticable. The cell is particularly well adapted to supply feeble currents, requiring a considerable E. M. F., especially when portability is desired.

Fig. 11, shows a single-fluid cell, consisting of a zinc-carbon couple in an exciting

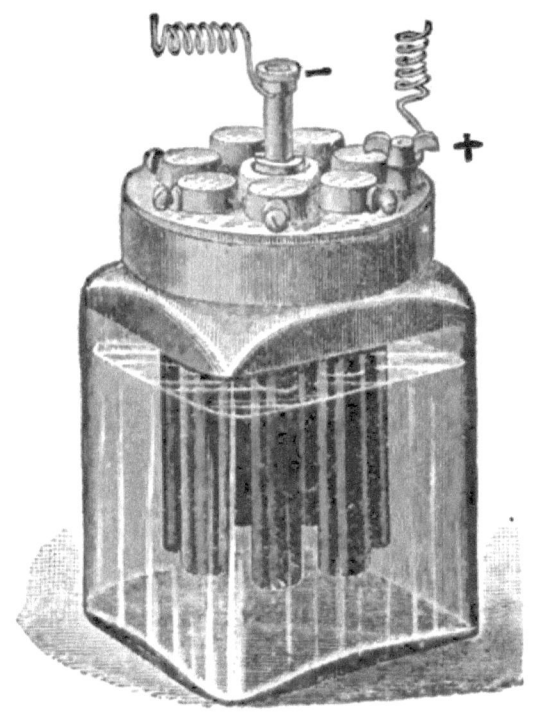

FIG. 11.—ZINC-CARBON CELL.

solution of sal-ammoniac in water. Here, as before, the terminal of the zinc plate forms the negative pole and that of the carbon plate, formed of a number of rods of carbon, the positive pole. This cell furnishes a strong current for a short time. Its advantage consists in the fact that it will supply a moderately strong current for a brief interval, and that it suffers very little chemical loss on open circuit. Its E. M. F. is about $1\frac{1}{3}$ volts. The disadvantage of the cell is that it polarizes considerably, so that it cannot continue to furnish current of any considerable strength for a long time.

Fig. 12, shows another form of a couple of zinc-carbon immersed in a solution called *electropoion fluid*, consisting of chromic acid and water, or of bichromate of potash, sulphuric acid and water. This cell is called the *Grenet* or *bichromate cell*. The advantage of the Grenet cell is that it is capable of supplying a fairly strong current for a short time, though longer than in the case of the simple carbon-zinc cell. Its disadvantage is that the chemical action

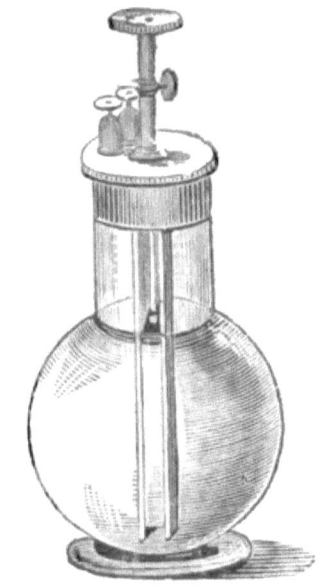

Fig. 12.—Grenet Plunge Cell.

continues even when the cell is on open circuit, so that the zinc becomes dissolved. Consequently, in practice provision has to be made in this cell, for raising the zinc plate from the solution when the battery is not in use.

Fig. 13, shows two different forms of

Fig. 13.—Edison-Lalande Cells.

Edison-Lalande cell. Here the couple is formed of plates of zinc and copper immersed in a solution of caustic soda, or potash, in water. Although this cell employs but a single liquid, yet, like the Leclanché cell, a special provision is made to prevent polarization. This is accomplished by placing a plate of compressed copper oxide in contact with the copper plate, so that the hydrogen, which tends to be liberated at the surface of the copper plate, is prevented from doing so by entering into combination with the oxygen of the oxide of copper. The Edison-Lalande cell possesses the advantage of furnishing powerful currents for a considerable length of time without sensible polarization, and also of suffering negligible local action or chemical loss on open circuit. Its disadvantage lies in its low E. M. F., which is only about two thirds of a volt, when at work.

Another form of Edison-Lalande cell is shown in Fig. 14. Here a large copper

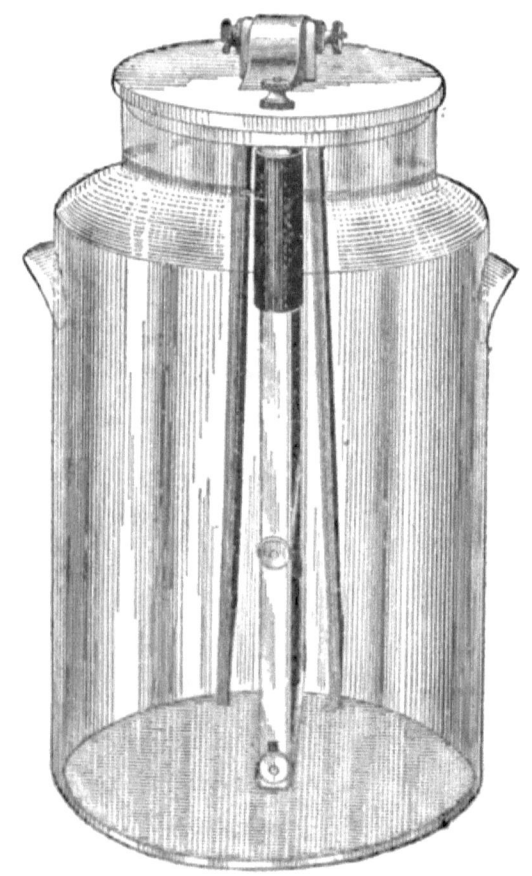

Fig. 14. Edison-Lalande Cautery Cell.

plate is placed between two zinc plates. The object of this form of cell is to provide powerful currents suitable for heating *electric cauteries;* i. e., metallic

Fig. 15.—Partz Gravity Cell.

wires or strips raised to a white heat by the passage of an electric current, and employed for removing diseased growths.

Fig. 15, shows a form of zinc-carbon cell, called the *Partz gravity cell*. This is a double-fluid, gravity cell, the fluids being a solution of common salt, or sulphate of magnesia, and a dense solution of a salt called sulpho-chromic salt, practically a salt which, dissolved in water, forms the electropoion solution before referred to. In order to charge the cell, the jar is first partly filled either with a solution of common salt or magnesium sulphate, and the sulpho-chromic salt added through the funnel shaped tube on the left-hand side of the cell, thus forming a dense solution surrounding the lower or carbon plate. The advantage of this cell is its high E. M. F., nearly two volts, and the strength of current it can supply. Its disadvantage is local action on open circuit.

Fig. 16, shows a form of cell called a

dry cell. This name is badly chosen, since, although the cell does not actually contain free liquid, yet its action is depend-

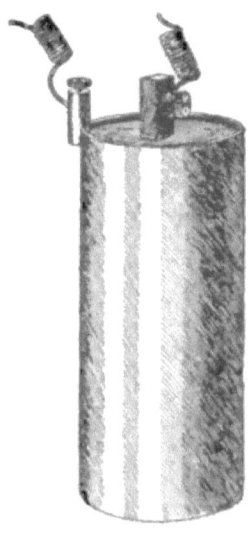

Fig. 16.—Form of Dry Cell.

ent upon the presence of a liquid electrolyte, in this case the liquid being absorbed either by a gelatinous substance, or by some pulverulent material. The advantage of a cell which contains no free liquid, is that the cell can be readily carried about without spilling any of the exciting liquid. The disadvantage of the cell is that the current it will supply is comparatively feeble, owing to its high resistance.

Dry cells give E. M. Fs. varying from two-third volt, to about two volts.

Where the circuit to be supplied is such that the E. M. F. furnished by a single cell is not capable of overcoming what is called the resistance of the circuit, it is necessary to connect a number of separate cells so that they may all supply their currents into the same circuit. A number of separate cells capable of acting as a single cell is called a *battery*, a term frequently incorrectly applied to a single cell.

A number of voltaic cells may be connected to form a battery, in a variety of ways, but at present we will discuss only one of such connections; namely, connection *in series*. The method adopted in this connection is shown in Fig. 17, where three Daniell gravity cells are shown, con-

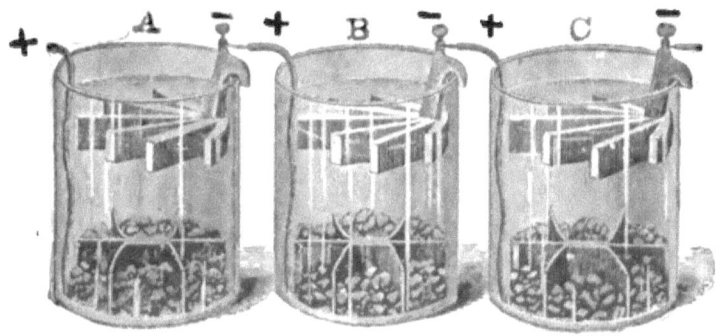

Fig. 17.—Series Connection of Three Daniell Gravity Cells.

nected in series. Here the negative pole of the cell A, is connected with the positive pole of the cell B; the negative pole of B, is connected with the positive pole of C, while the free positive and the free negative poles of A and C, respectively, form the terminals of the battery. In the case of the series combination of voltaic cells, the E. M. F. of the series is equal to the sum of the E. M. Fs. of the separate

Fig. 18.—Edison-Lalande Battery.

cells composing it. Consequently, the battery shown in Fig. 17, will have an E. M. F. of approximately three times 1.05 volts.

Fig. 18, shows a battery of two series-connected, Edison-Lalande cells. Fig. 19, shows a battery of the general type shown in Fig. 12. Here each cell is formed of a number of plates of zinc and carbon. When the battery is desired for use, the

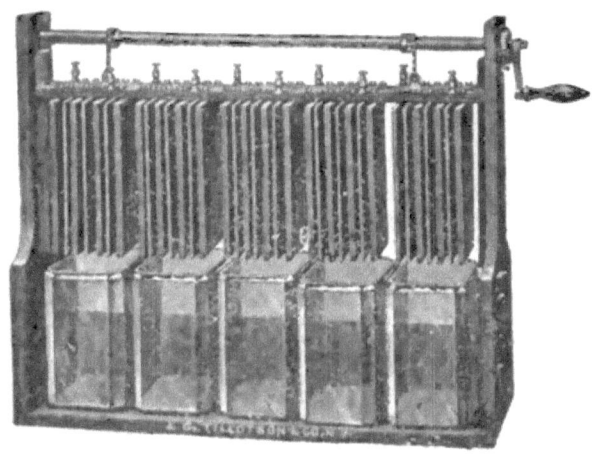

Fig. 19.—Plunge Battery.

handle is turned so that the plates are let down into the jars filled with the exciting liquid. When out of use, the plates are raised from the jars, in order to preserve them from corrosion, or local action.

ELECTRO-THERAPEUTICS. 53

Should the plates not be removed from the exciting liquid, the zincs will rapidly be corroded.

Fig. 20, shows a battery of 50 series-connected, silver-chloride cells, so arranged

Fig. 20.—Silver-Chloride Battery of Fifty Cells.

that any number from one up to 50 can be connected to the main terminals.

A voltaic cell will continue to furnish current to the circuit connected with it, as long as it contains any positive metal to be dissolved, and any electrolyte to dissolve it. As soon as either the electrolyte or the positive plate is consumed, the cell ceases to give current until it is furnished with another plate or with more electrolyte, or both. In the form of voltaic cell called the *storage cell*, or *secondary cell*, to distinguish it from the ordinary voltaic or *primary cell*, the selection of the elements of the couple and the electrolyte are such, that after the cell is completely *exhausted* or *run down*, it is capable of being *restored* or *charged*, and of again being brought into a condition ready for action by the passage of an electric current through it in the opposite direction to that of the current which it furnishes when charged. The passage of this *charging current* has the effect of producing a series of decompositions, which practically restore the condition both of the elements of the couple and of the exciting liquid or electrolyte.

Secondary or storage cells are made in a variety of forms, but those in common use, consist practically, when charged, of a voltaic couple of porous lead and lead peroxide, immersed in an electrolyte of dilute sulphuric acid. The lead peroxide forms the positive plate, and the metallic finely divided, porous lead, the negative plate.

In most forms of storage battery, the substances forming the positive and negative elements are packed in perforations in a supporting plate or *grid*, made of an alloy of lead containing a small percentage

Fig. 21.—Plate of Chloride Storage Battery.

of antimony. The object of the antimony is to prevent the action of the sulphuric acid on the lead in the grid during charging. Fig. 21, shows a plate of a well-known form of storage cell, called the

ELECTRO-THERAPEUTICS 57

chloride storage cell. This plate consists, as shown, of a lead-antimony grid, contain-

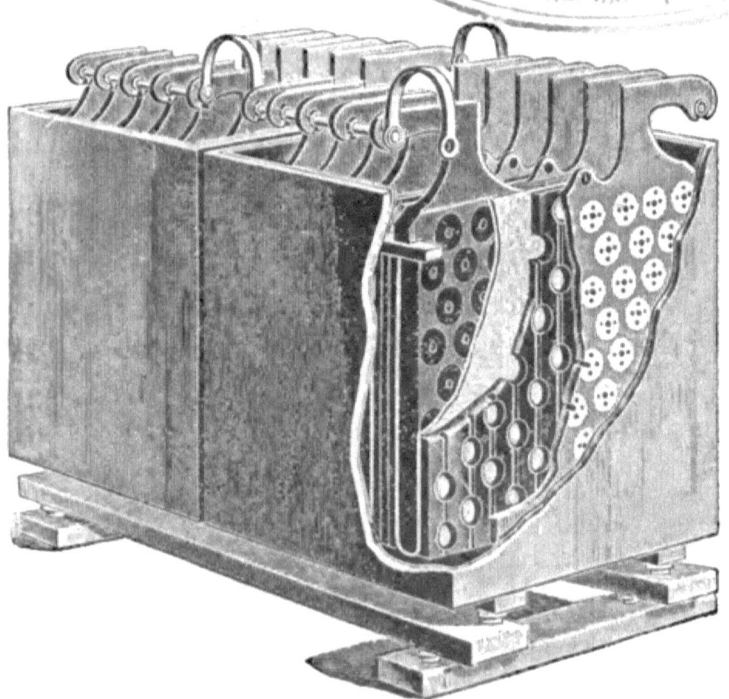

Fig. 22.—Interior View of Chloride Storage Battery.

ing pastels or discs of metallic lead or lead peroxide, according to the nature of the plate. A number of such plates are generally grouped together, to form a single storage cell, the connections being such that all the positive plates are con-

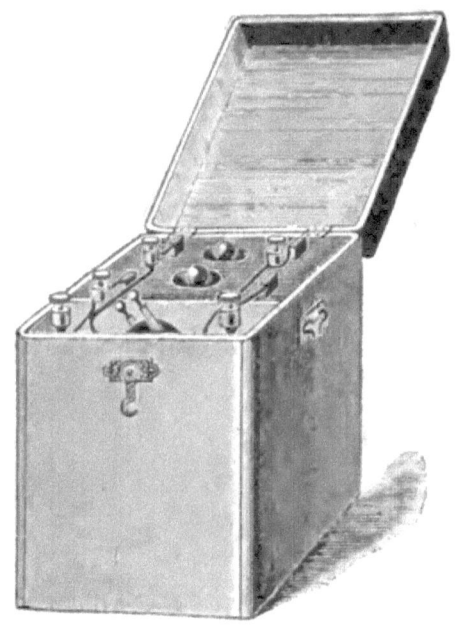

Fig. 23.—Portable Chloride Storage Battery.

nected together to form a single positive plate, and all the negative plates are similarly connected to form a single negative plate, the whole being immersed in a jar containing a solution of sulphuric acid in water. In Fig. 22, two of such storage cells are connected together to form a *storage battery*. A portable form of chloride storage battery suitable for medical purposes is shown in Fig. 23.

Another form of storage cell, called the *Julien Cell*, is shown in Fig. 24. The advantage of storage cells is that they are capable of supplying a very powerful current, and that the E. M. F. of each cell is two volts. Their disadvantage is that they require to be periodically charged.

Storage cells are generally charged by connecting them with the terminals of a dynamo-electric machine, the positive terminal being connected to the positive terminal of the dynamo. The dynamo may be provided especially for the purpose, or the charging current may be

Fig. 24.—Julien Storage Cell.

obtained from the mains supplying electric incandescent lamps. Sometimes, how-

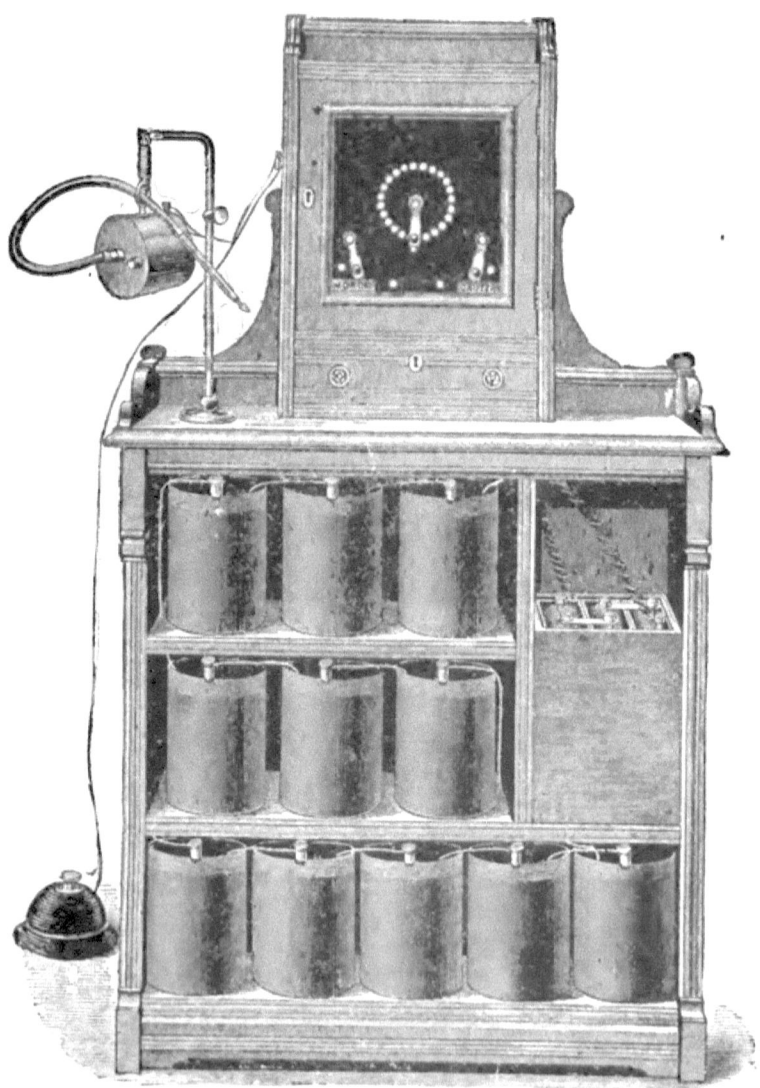

Fig. 25.—Cabinet Containing Eleven Primary Voltaic Cells for Charge of Two Storage Cells.

ever, a primary battery is employed for this purpose. For example, as shown in Fig. 25, 11 gravity cells are connected in a battery so as to charge two storage cells. The current supplied by the primary battery is a feeble but steady current, and the storage cells when charged by the action of this current, may be many times stronger for a correspondingly shorter length of time.

CHAPTER III.

ELECTRIC RESISTANCE.

WE have pointed out, in the preceding chapter, that it is not electricity which an electric source primarily produces, but an electromotive force, or a force capable of producing a flow or current of electricity, when the terminals or poles of the source are suitably connected by a conducting path or circuit. We have also seen that electromotive force is measured in units called volts, and that an ordinary bluestone gravity cell produces, under ordinary conditions, an E. M. F. of a little more than one volt. When the terminals of a single voltaic cell of this type are connected with a circuit, the electric current which will flow through the circuit, that is, the amount of electricity which will pass through it per second, will depend upon another quantity called the *resistance* of the circuit. Electric resistance is that which opposes the flow of electricity through a circuit.

Resistance is measured in *units of resistance* called *ohms*, after Dr. Ohm, who first pointed out the law which governs the flow of electricity through a circuit. The resistance of a circuit depends upon its length, its area of cross-section, and the material of which it is composed. A short, thick circuit, of some good conducting material like copper, will have a small resistance; a long, thin circuit of some poor conducting material, like wet string, will have a high resistance. The greater the length of a conductor, the greater will be its resistance. Thus; if we double the length of any uniform metallic wire, we double its resistance; or, if we halve the length of the wire, we halve its resistance. The greater the area of cross-section of a conductor the less its resistance, so that if the area of cross-section of a wire be doubled, retaining the same length, its resistance will be halved.

The ohm is the resistance of a definite length of a definite material, having a definite area of cross-section. The ohm is defined as the resistance of a column of pure mercury having a length of 106.3 cms., and a cross-sectional area of one square millimetre, at the temperature of melting ice, or 0° C. The actual definition requires, not that the cross-sectional area should be one square millimetre, but that the weight of the column of mercury whose length is 106.3 centimetres, should be 14.4521 grammes, but this is equivalent, in a uniform column, to a cross-section of one square millimetre. In defining the length and the area of cross-section of the column of mercury which represents the ohm, it is necessary to specify the temperature, since the resistance of metallic bodies increases with an increase in temperature. The ohm may also be roughly stated as being the resistance offered by two miles of ordinary copper trolley wire, or by one foot of copper wire No. 40 A. W. G. (American Wire Gauge) having a diameter of 0.003145 in., at 45° F.

In order to compare the relative resistances of wires of different materials, having the same length and area of cross-section, as well as for the purpose of being able to calculate the resistance of a given length and area of cross-section of any material, it is usual to consider the *specific resistance* or *resistivity* which bodies offer to the passage of electric currents. The resistivity of a substance is numerically equal to the resistance offered by a wire of such substance having unit length and unit cross-section. Any units of length and cross-section might be adopted for this purpose, but the units actually adopted are the centimetre, for the unit of length, and the square centimetre for the unit of cross-sectional area. The resistivity of a substance, therefore, is numerically equal to the resistance offered by a wire of the substance, one centimetre long and one square centimetre in area of cross-section. In the case of metals, the resistivity is always a very small fraction of an ohm, and is, in fact, usually expressed in *microhms;* i. e., in millionths of an ohm. In the case of many liquids, the resistivity is conveniently expressed in ohms, but in the case of materials which possess very poor conductive power, generally called *insulators,* the resistivity becomes enormously great and is more conveniently expressed in *megohms,* or millions of ohms, in *begohms,* or billions of ohms, i. e., thousands of millions, or in *tregohms,* or trillions of ohms, i. e., millions of millions.

The following is a table of resistivities of various substances:

Silver, annealed,	. . 1.53 microhms,	at 0° C.
Copper, "	. . . 1.594 "	"
Iron, "	. . . 9.687 "	"
Mercury,	. . . . 94.84 "	"
Platinum,	. . . . 9.03 "	"
Pure water,	. about 3.75 megohms,	at 10° C.
Tap water, .	. " 200,000 ohms,	at 10° C.
Hard rubber,	. . . 28,000 tregohms.	
Porcelain,	. . . . 540,000 "	

In order to show the use of this table, suppose it to be required to calculate the resistance of a wire of platinum one foot long (30.48 cms.), and $\frac{1}{100}$ of an inch in diameter; i. e., having a cross-sectional area of $\frac{1}{12,730}$ square inch=0.0005067 square centimetre.

Looking at the table of resistivities we find for platinum the value 9.03 microhms, and the resistance of the wire will, therefore, be $\frac{9.03 \times 30.48}{0.0005067} = 543,200$ microhms= 0.5432 ohm. Here we multiply the resistivity by the length because the longer the wire, the greater will be its resistance. If the wire be 30.48 centimetres long, and one square centimetre in area of cross-section, it would have a resistance of

9.03×30.48 microhms. We divide by the area, because the greater the area, the less the resistance. The resistance of a platinum wire, one centimetre long, and 0.0005067 square centimetre in cross-sectional area, would have a resistance of

$$\frac{9.03}{0.0005067} = 17,820 \text{ microhms} = 0.01782 \text{ ohm}.$$

The resistance of any metal can, theoretically, at least, be computed in the same manner, but slight impurities in the material are liable to affect markedly its resistivity, and consequently its resistance, so that, except in the case of very nearly pure metals, the resistances of computation are not very reliable. The effect of impurities is always to increase the resistivity, and, therefore, the resistance.

It will be observed that most of the resistivities are given at the temperature

0° C. Ordinarily, the effect of temperature is to increase the resistivity of all metallic substances. This effect is nearly the same for all pure metals, and is, roughly, 4-10ths of one per cent. per degree centigrade increase of temperature above zero centigrade. In the case of liquids, and of non-metallic substances generally, the resistivity diminishes as the temperature rises. Carbon behaves in this respect like an insulator, rather than a conductor. Temperature has a marked influence in diminishing the resistivity upon the best insulators.

It may be seen from the table, that the resistivity of pure water is very high; namely 3.75 megohms, and it is believed by some, that if water could be obtained in absolute purity it would not conduct, or that its resistivity would be indefinitely great. Very slight degrees of impurity suffice, however, to greatly reduce the resistivity, and the addition of a soluble salt reduces it to a few ohms. Thus the resistivity of a strong solution of zinc sul-

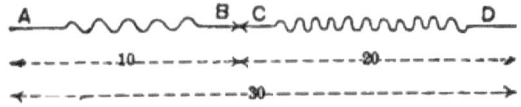

Fig. 26.—Connection of Resistances in Series.

phate in water is about 30 ohms at ordinary temperatures.

If a wire AB, of 10 ohms resistance, be connected with a second wire CD, of 20 ohms resistance, as shown in Fig. 26, in such a manner that the current first passes through one and then through the other, they are said to be *connected in series*, and the total resistance of the series AD, will be $20+10=30$ ohms.

If two conductors, *AB* and *CD*, Fig. 27, each of 10 ohms resistance, be *connected in parallel*, as shown, so that the current divides between them, as indicated

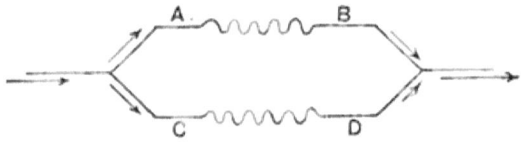

Fig. 27.—Connection of Resistances in Parallel.

by the arrows, the *joint resistance* of the pair will be $\frac{10}{2}=5$ ohms. Similarly, if three such wires be connected in parallel, their joint resistance would be $\frac{10}{3}=3.333$ ohms, and so on, for any number of parallel wires.

The resistance of any instrument wound with wire, such, for example, as a telephone, depends upon the length and cross-section of the insulated wire employed, as well as on the character of the wire itself. For a given size of coil; *i. e.*, a given volume of winding space, the resistance increases rapidly as the diameter of the wire employed to fill that space is reduced. Neglecting the variation of the insulating thickness of the wire, and its effects, the resistance will increase inversely as the fourth power of the diameter; that is to say, if we halve the diameter, we shall increase the resistance of the winding approximately 16 times ($2^4 = 2 \times 2 \times 2 \times 2$) = 16.

If one Daniell gravity cell, represented in Fig. 28, be connected through 200 feet of No. 25 A. W. G. copper wire, to a telephone, the resistance of the telephone is say 50 ohms, the resistance of the wire 6.5 ohms, and the resistance of the cell, say 5 ohms. The total resistance of the circuit will, therefore, be 50+6.5+5=61.5 ohms.

The electric resistance of any organic substance, such as moist flesh, is very difficult to determine. As soon as flesh is dried, it becomes a non-conductor, and it is,

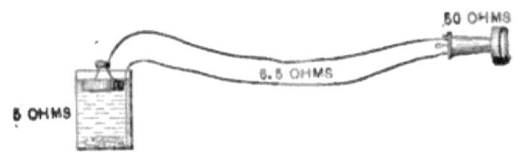

Fig. 28.—Resistances in a Circuit Consisting of a Voltaic Cell, Telephone, and Connecting Wires.

therefore, evident that the conducting properties of the mass are due almost entirely to the conducting power of the liquids contained in it. These liquids may be regarded as forming physically a continuous mass, although, in reality, the mass is divided by numerous porous partition walls, or septa. The effect of these division walls is to increase greatly the resistance offered by the liquids.

For the same reason, the resistance of the human body, or of any portion of the body, is a very complex quantity, and varies from time to time. It also varies with the nature of the area of the contact surfaces between which the measurement is taken. Thus if a bare copper wire be grasped in each hand, the resistance of the body, as measured between the two copper wires, may be 100,000 ohms, or more, depending largely upon the dryness of the skin. If now, instead of holding the two wires, the wires be connected to metallic electrodes, such as those shown in Fig. 29, the resistance apparently offered by the body, as measured between them, will, perhaps, be only 30,000 ohms, for the reason that the area of cross-section of skin, through which the current enters and leaves the body, has been greatly increased; *i. e.*, the area of cross-section of the conductor in the skin itself has been greatly increased. If now, the

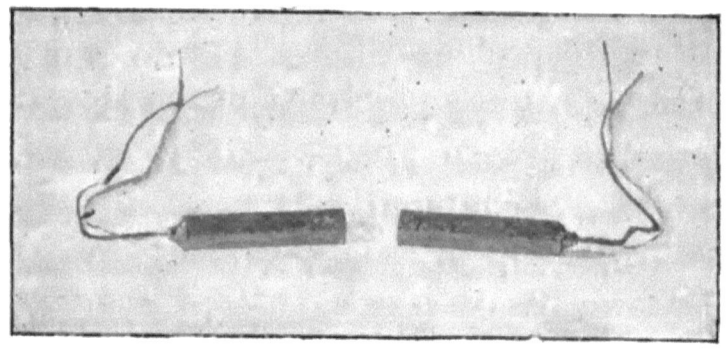

Fig. 29.—Metallic Handle Electrodes.

same electrodes be held in the hands, while wrapped in absorbent cotton, thoroughly wetted by, say salt water, the resistance of the body, as measured between them, will be, perhaps, 5,000 ohms, owing to the fact that the skin has become thoroughly moistened by contact with the saline solution, and has thereby become a better conductor, or has had its resistance lowered. Again, if the hands be dipped to the wrists in jars containing salt and water, or dilute caustic soda, so that the entire surface of the hands is brought into contact with the liquid, the resistance of the body, as measured between the two jars, will, probably, be only 1,000 ohms. If, finally, the hands be dipped still deeper in tall jars containing the same solution, the resistance will still tend to slightly diminish, owing to the greater area of skin offering passage to the current, and the reduced effective length of the circuit through the body.

The principal resistance which the body offers to the passage of an electric current is that of the skin, owing, not only to the nature of its substance, but also to the fact that under ordinary circumstances it is dry. The strength of current, therefore, that will pass through the human body, in the event of an accidental contact with an electric circuit, such for example as a trolley wire, will depend markedly on the nature of the contact, as well as on the electromotive force in the parts of the circuit in contact; in other words, the current received will depend not only on *what* is touched, but also on *how* it is touched.

CHAPTER IV.

ELECTRIC CURRENT.

WHEN we speak of an electric current flowing through a circuit we mean the rate at which electricity passes through the circuit, and this rate can be expressed by the amount of electricity which passes through the circuit in a given time, say in one second, just as the flow of water through a pipe can be expressed by the quantity of water which passes in a given time. The *unit quantity of electricity* is called the *coulomb*. A little consideration will show that in the case of liquid flowing through a pipe, since the entire mass of the liquid in the pipe is in motion, the quantity which passes any cross-section of the pipe, in a given time, must be the same as that which passes through any other cross-section in the same time, since, otherwise, there would tend to be a surplus in some parts of the pipe, and a deficit in others. The same is true concerning an electric flow or current. The amount of electricity passing any point of the circuit being always rigorously equal to the quantity which passes any other point in the same time. We distinguish between the quantity of electricity and the rate at which it flows, just as we distinguish between a gallon of water, and the rate of flow of a gallon-per-second. The electric *unit rate of flow*, or unit of current strength, is called the *ampere*, and is equal to one coulomb-per-second.

The ampere is the unit rate of flow invariably employed in all practical applications of electricity. Thus, an electric incandescent lamp may require for its operation a current strength of from 0.25 to 10 amperes, according to its dimensions, and the amount of light it is designed to produce. In some applications of electricity, as, for example, in electric welding, or metal working, enormous currents have to be employed, 50,000 amperes being sometimes required. In the application of electricity in electro-therapeutics, the current strength is always a small fraction of an ampere, and is generally measured in *milliamperes*, or thousandths of an ampere, for the reason that a current strength of one ampere would be dangerously great. In the application of the death penalty by electricity, as practiced in the State of New York, the alternating current strength passed through the body of the criminal is usually 7 or 8 amperes.

When an electric current is sent through a metallic solution, such, for example, as an aqueous solution of copper sulphate, the passage of the current is attended by a decomposition of the salt, metallic copper being deposited on the terminal or electrode connected with the negative pole of the battery, and an acid substance appearing at the electrode connected with the positive pole, and entering into combination with it. If such combination be chemically impossible this substance will be liberated at the electrode. Decomposition by electricity is called *electrolysis*. The amount of electrolytic decomposition, that is, the amount of saline substance decomposed, will depend upon the quantity of electricity which passes, as well as upon the nature of the substance itself. An ampere may, therefore, be defined by the amount of chemical decomposition which it can effect in a given time. Thus, since an ampere is a current of one coulomb-per-second, and each coulomb of electricity passing through a solution of silver deposits 1.118 milligrammes of silver, a current strength of one ampere produces a deposit of 1.118 milligrammes of silver per second.

When the resistance of a circuit is known in ohms, and the electromotive force applied to the circuit is known in volts, the strength of current, which passes through the circuit in amperes, can be readily calculated by a law known as *Ohm's law*. Ohm's law is generally represented by the equation,

$$\text{Amperes} = \frac{\text{Volts}}{\text{Ohms}}.$$

Suppose, for example, a circuit having a resistance of 10 ohms has an E. M. F. of

5 volts acting on it; then the current which flows through the circuit will be $\frac{5}{10}=\frac{1}{2}$ ampere. The practical electric units of E. M. F., resistance and current strength may, therefore, be defined, in terms of Ohm's law, as follows;

A volt is such a unit of E. M. F. as will produce a current of one ampere in a circuit whose electric resistance is one ohm.

An ohm is such a unit of electric resistance, as will limit the flow of electricity to a current of one ampere, when under an E. M. F. of one volt.

The ampere is the rate of flow of current which will pass through a circuit whose resistance is one ohm, under an E. M. F. of one volt.

The above definitions, although convenient for defining electric units in terms of one another, are not generally employed.

The following examples of the application of Ohm's law will show its importance. Required the E. M. F. necessary to produce a current strength of 10 milliamperes through the human body, its resistance being, under given conditions, 5,000 ohms.

10 milliamperes $= \dfrac{10}{1,000} = \dfrac{1}{100}$ ampere, and the E. M. F. which divided by 5,000 ohms gives $\dfrac{1}{100} = 50$ volts, the voltage required.

A battery of fifty silver-chloride cells, each having an E. M. F. of 1.05 volts, and an internal resistance of 10 ohms per cell; it is required to know whether a single cell or the whole battery of fifty cells in series will give the greater current strength through a short stout piece of copper wire.

The resistance of the short piece of copper wire being negligibly small compared with the resistance of a single cell, we may omit it altogether in the calculation. The current strength from one cell will, therefore, be $\frac{1.05}{10} = 0.105$ ampere $= 105$ milliamperes. With fifty cells we have 50 times as much E. M. F. and also fifty times as much resistance, and therefore, by Ohm's law, the current strength will be $\frac{52.5}{500} = 0.105$ ampere $= 105$ milliamperes, or the same as before, so that it is evident that through a negligibly small external resistance, or, as it is called, on *short circuit*, there is no advantage in adding similar cells in series, since although each cell adds its E. M. F. to the circuit, it also adds a proportional amount of resistance.

If the silver-chloride battery of the preceding paragraph, is employed to send a current through an external resistance of 1,000 ohms, what will be the current strength with one cell and with fifty cells?

With one cell, the total resistance will be $1,000+10 = 1,010$ ohms, and the E. M. F. 1.05 volts, so that the current strength will be $\dfrac{1.05}{1,010} = 0.00104 = 1.04$ milliampere.

With fifty cells, the total resistance will be $1,000+500 = 1,500$ ohms, and the E. M. F. $1.05 \times 50 = 52.5$ volts. The current strength will, therefore, be $\dfrac{52.5}{1,500} = 0.035 = 35$ milliamperes.

As another example, let us suppose that two Gravity-Daniell cells have each an E. M. F. of 1.05 volts, and a resistance of 4 ohms. Find the current strength which they can send through a resistance of one ohm externally; (*a*) When connected singly; (*b*) In series; (*c*) In parallel.

(*a*) $\dfrac{1.05}{4+1} = \dfrac{1.05}{5} = 0.21 = 0.21$ ampere = 210 milliamperes.

(*b*) In series, $\dfrac{1.05 \times 2}{4 \times 2 + 1} = \dfrac{2.1}{9} = 0.233 =$ 233 milliamperes.

(*c*) In parallel. Here the positive pole of one cell is connected to the positive pole of the other, and the negative pole of one cell, connected to the negative pole of the other. The E. M. F. of the combination will be that of a single cell, but the resistance of the combination will be that of two equal resistances in parallel, or one half that of either; namely, 2 ohms, the current strength will, therefore, be $\frac{1.05}{2+1} = \frac{1.05}{3} = 0.35$ ampere = 350 milliamperes.

An instrument for measuring the strength of electric currents is called an *amperemeter*, or *ammeter*. For electrotherapeutic purposes, since the current strength to be measured is usually expressed in milliamperes, the instrument is frequently called a *milliammeter*.

Milliammeters are made in a variety of forms. In nearly all cases, however, an index or pointer is moved over a graduated scale by the force exerted between a coil of wire carrying the current to be measured, and a magnet in its vicinity. This movement is due to the fact that a wire carrying an electric current becomes temporarily invested with magnetic properties.

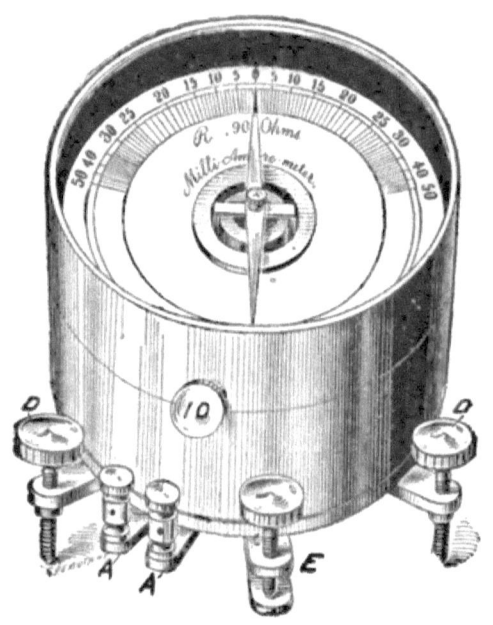

Fig. 30.—Form of Milliammeter.

A simple form of milliammeter is shown in Fig. 30. In this a pair of coils of wire, situated beneath the horizontal face of the instrument, become magnetized by the passage of the current to be measured, and deflect a magnetic needle, in the shape of a split bell, from the position it assumes under the influence of the earth's field at the place where the measurement is made.

Fig. 31.—Vertical Form of Milliammeter.

The scale is marked directly in milliamperes.

Another form of instrument is shown in Fig. 31. Here a vertical needle is deflected by the attraction of a coil of wire upon a magnetized needle placed inside the instrument.

FIG. 32.—FORM OF MILLIAMMETER.

Still another form of instrument, differing only in constructive details from those above described, is shown in Fig. 32.

While the preceding instruments are simple in their construction, yet they are all liable to have their indications attended by changes of magnetic force occurring in their neighborhood, and even when all magnets are carefully removed from their vicinity, their indications are often at-

Fig. 33.—Weston Milliammeter.

tended by appreciable error, due to some difference between the strength of the earth's magnetic force at the place where the instrument is employed, and the place where it was originally calibrated.

A form of instrument which is practically free from the earth's magnet influ-

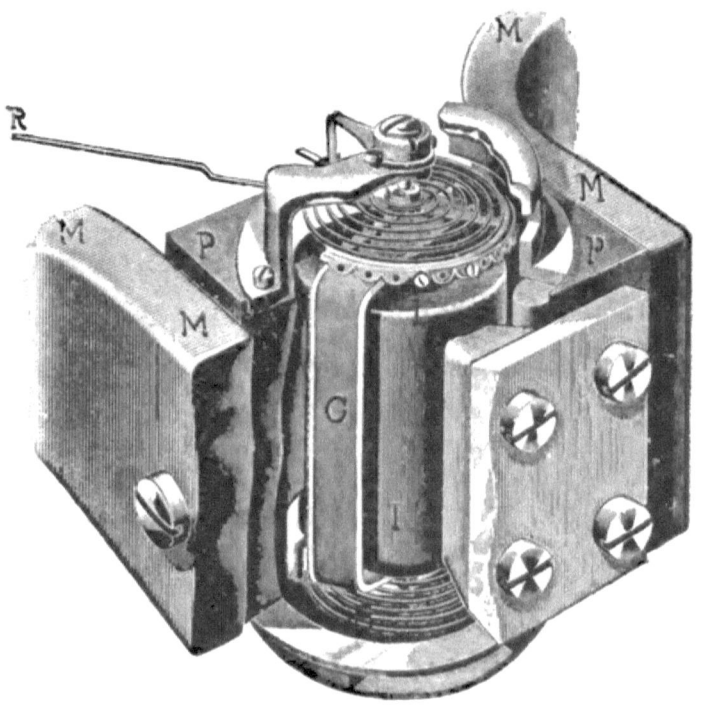

Fig. 34.—Working Parts of Weston Ammeter.

ence is shown in Fig 33. This freedom is owing to the fact that the instrument has no iron moving parts. The principal working parts are shown in Fig. 34. A horse-shoe permanent magnet MM, MM, only partly visible in the figure, has soft iron projections, or pole-pieces P, P, secured to its poles. These projections are shaped so as to enclose a cylindrical space. At the centre of this space is supported a soft iron, solid cylindrical core I I. Between this core and the pole-pieces, there remains a narrow annular gap, or space, which is permeated by the magnetic flux from the permanent magnet. In this space a delicately supported coil C, of insulated copper wire, is free to move. Coil springs S, S, carry the current to be measured into and out of the coil C. So long as there is no current through the coil, it is unaffected by the magnetic field in which it is placed, and the pointer remains at the zero point. When, however, a current passes through the coil, the electromagnetic action of this current upon the magnetic field, causes a mechanical force to be exerted upon the coil, deflecting it against the spiral springs S, S, in such a manner that the pointer R, moves over a scale beneath it through a distance, which is almost directly porportional to the current strength. The permanent magnet's field is so very much more powerful than the earth's magnetic field, that the influence of the latter upon the coil is negligible by comparison. The accuracy of the instrument depends upon the degree of permanence with which the magnetism in the horseshoe magnet is retained. The instrument has the disadvantage that it will not read in either direction, so that if the current passing through the instrument has the wrong direction, the wires attached to the instrument must be reversed.

In several of the instruments just described, means are provided for varying their sensibility and the range of their indications. Thus, in Figs. 29 and 30, a screw button is seen projecting from the face of the apparatus, marked 10. If this screw be pressed forward, until it abuts strongly against its stop, a coil of wire will be brought into connection with the terminals in such a manner that 9-10ths of the current passing through the instrument will pass through this wire, and only 1-10th will pass through the measuring coils. The instrument under these conditions is said to be *shunted*, or to have a shunt applied to it whose power is 10. The readings of the instrument in milliamperes must now be multiplied by 10, in order to obtain the actual current strengths. In Fig. 33, three terminals are shown and the instrument is provided

with two sets of graduations on its scale.
When the right hand and the corner left
hand terminals are used, the lower scale is
brought into use, by which the instrument
reads up to 10 milliamperes only, but by
using the right hand, and the inner left
hand terminals, the upper scale is utilized,
by which currents up to 500 milliamperes
can be measured.

Instruments of the preceding types are
sufficiently sensitive for all the ordinary
requirements of electro-therapeutic applications. When, however, physiological researches have to be made, in which very
feeble electric currents are measured, it is
necessary to use a mirror galvanometer, as
a form of ammeter. Such an instrument
is made in a variety of forms, one of the
simplest of which is illustrated in Fig. 35.
A circular coil or spool of fine insulated

copper wire is mounted upon a tripod frame, and a small magnetized needle

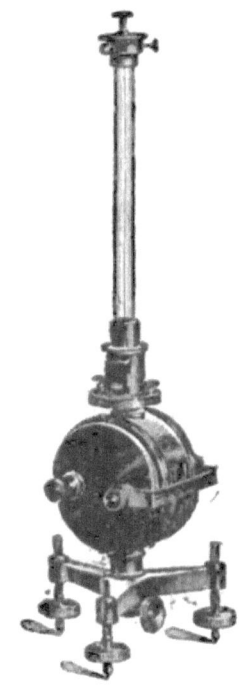

Fig. 35.—Mirror Galvanometer.

is suspended by a fibre of silk at the centre of the coil. A small glass mirror is attached to the suspension in such a manner that any deflection of the little magnetized needle will cause an angular deflection of the mirror. Facing the instrument is a lamp and scale, shown in Fig. 36. The lamp, whose glass

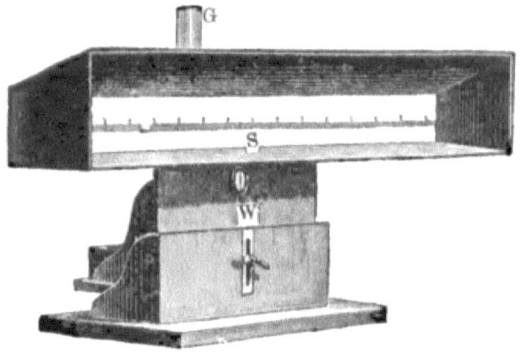

FIG. 36.—MIRROR GALVANOMETER SCALE.

chimney only is seen at G, throws a beam of light through the window W, on the mirror suspended in the galvanometer. The latter reflects the beam on the scale S. Since a deflection of the mirror through an angle of 45°, would be sufficient to deflect the beam through a right angle, or 90°, and, therefore, to send the beam to the end of an indefinitely long scale, it is evident, that a very small angular deviation of the magnet under the influence of the current to be measured passing through the coil, will suffice to produce a marked displacement of the spot of light along the scale S. A current in one direction will deflect the beam to the right, and a current in the opposite direction, to the left. Such an apparatus has commonly a resistance of 500 ohms, and a current of one millionth of an ampere, on one *micro-ampere*, will deflect the beam through a distance of a millimetre on a scale one metre distant.

In researches of very great delicacy, where exceedingly feeble currents have to be observed, special very sensitive mirror galvanometers are employed. One of these

ELECTRO-THERAPEUTICS. 103

is shown in Fig. 37. Here four sets of coils, one above another, act on four little mag-

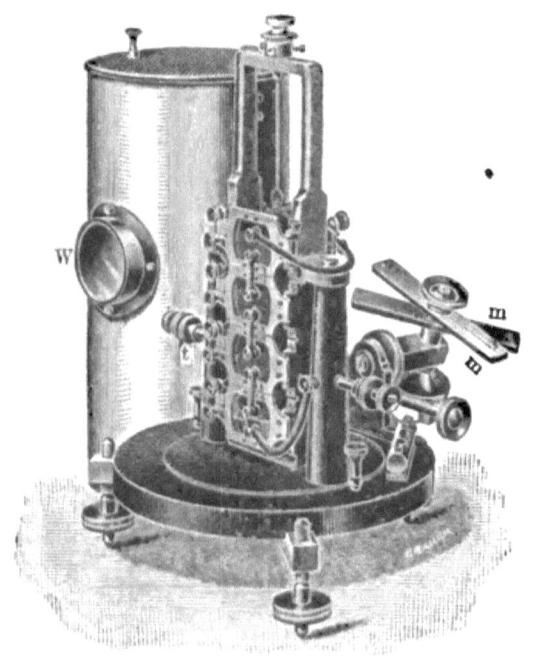

Fig. 37.—Sensitive Mirror Galvanometer.

netic needles situated at their respective centres. A single mirror, attached to the upper part of the suspension, reflects its beam of light through the window *W*. The terminals of the coils are brought to the connecting posts *t, t*. *m, m*, are two *controlling magnets* employed for bringing the magnetic needles back to the same position after the application of the current to be measured. Such an instrument has a resistance of about 15,000 ohms, and has a sensibility such that one billionth of an ampere, or one *bicro-ampere, i. e.* one thousand millionth ampere, will produce a deflection of 15 millimetres on a scale distant one metre. Except for delicate work, and very feeble currents, the Thomson galvanometer is undesirable, as the values of its indications have usually to be converted into amperes by careful measurements and computations.

A form of galvanometer, very convenient when the greatest sensibility is not required, is shown in Fig. 38, and is called the D'Arsonval galvanometer. Here a

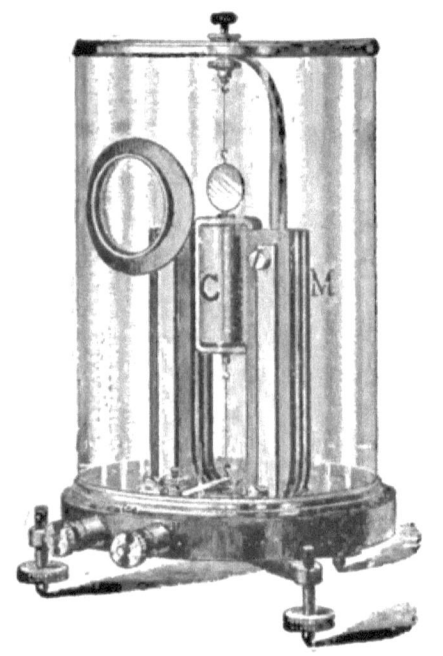

Fig. 38.—D'Arsonval Galvanometer.

coil of insulated wire C, is suspended between the poles of a permanent magnet M, and by means of the attached mirror, the deflection of this coil can be observed. It is evident that while in the preceding mirror galvanometers, **the coil is fixed and the** magnet is movable, in this instrument the magnet is fixed and **the coil is movable.**

CHAPTER V.

VARIETIES OF ELECTROMOTIVE FORCE.

THE voltaic or primary cell, and the secondary cell already described, will produce an E. M. F. which, so long as the chemicals remain unchanged, does not vary in strength. Such an E. M. F. is, therefore, called a *continuous E. M. F.* A continuous E. M. F. is also produced by a variety of other electric sources, such, for example, as a *continuous-current dynamo*, which, so long as its speed of rotation remains the same, produces an E. M. F. which is practically continuous.

Fig. 39, represents graphically a continuous E. M. F. The straight line *AB*, is drawn parallel to the base *OS*, at a distance representing 1.1 volts. Time is measured along the base *OS*, and the fact that the line *AB*, remains parallel to the

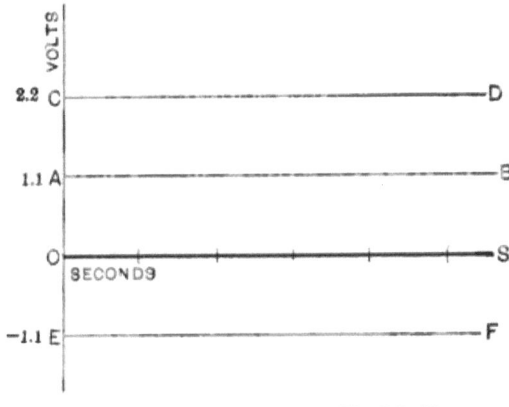

Fig. 39.—Continuous E. M. F.

base, represents the constancy of the E. M. F., which might be that of a single Daniell cell. Two such cells, connected in series, would produce a continuous E. M. F. of 2.2 volts, represented by the straight line *CD*, twice as far above the line *OS*, as the line *AB*.

An E. M. F. possesses direction, as well as magnitude; that is to say, it may tend to send a current through a circuit in one direction or in the opposite direction. All E. M. F.'s that tend to send the current in one direction may be regarded as positive, and all tending to send the current in the opposite direction, as negative. *Positive E. M. F.'s* are represented graphically by distances above the line *OS*, and *negative E. M. F.'s*, by distances below. Thus, in Fig. 39, the line *EF*, would indicate a negative E. M. F. of 1.1 volts, or an E. M. F. oppositely directed to that of the line *AB*.

Fig. 40, shows the E. M. F. produced by a continuous-current dynamo. Here the line *AB*, is parallel to the base as before, but instead of being straight, is a fine, wavy line. These little waves represent variations in the amount of E. M. F. produced every time that the bar in the commutator passes underneath the collecting brush. These wavelets exist in the E. M. F. of every continuous-current

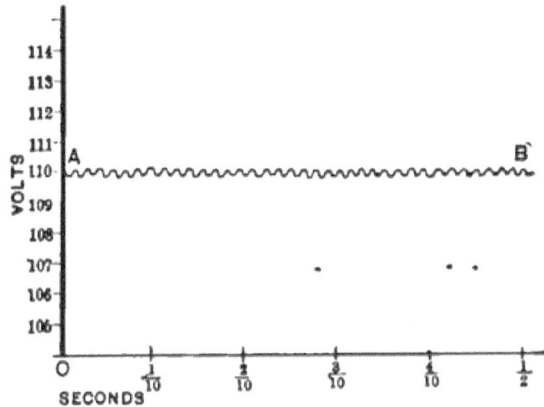

Fig. 40.—Type of E. M. F. Produced by a Continuous-Current Dynamo.

dynamo. When they are very marked, as represented in Fig. 41, the E. M. F. is said to be *pulsatory*. Such E. M. F.'s are produced by some continuous-current generators, usually for supplying arc lamps. It is evident, that at different times the E. M. F. varies considerably in its magnitude, but never changes direction, the line AB, being always on one side of the zero line OS; that is to say, it always has the same direc-

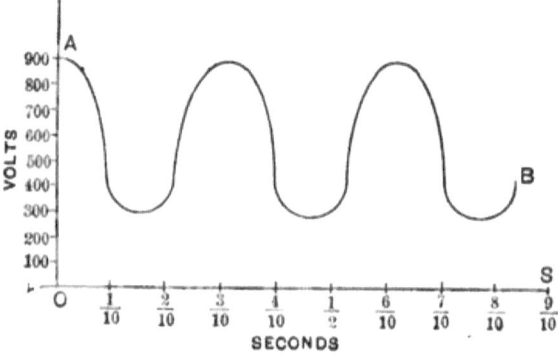

FIG. 41.—PULSATORY E. M. F.

tion in the circuit, just as though a battery of voltaic cells were employed to send current through a circuit, and that at intervals, a certain number of these cells were cut out and re-introduced.

When the waves start each time from the zero line, the E. M. F. is said to be *intermittent*. Fig. 42, shows that, at certain intervals, an E. M. F. exists in the circuit in one direction, and that at intervening

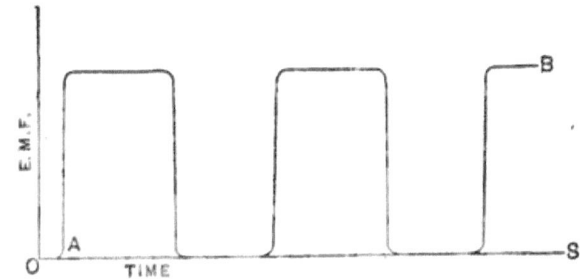

Fig. 42.—Intermittent E. M. F. Undirectional.

intervals there is no E. M. F. The intermittent E. M. F. can be obtained by connecting a continuous E. M. F., say a voltaic battery, to a wheel interrupter, in such a manner that the E. M. F. will be periodically cut off and applied. In all cases, although the strength of the E. M. F. varies at different times, yet at no time does it change direction, so that the curved line lies wholly above the base line. When an E. M. F. changes direction, as well as magnitude, it becomes

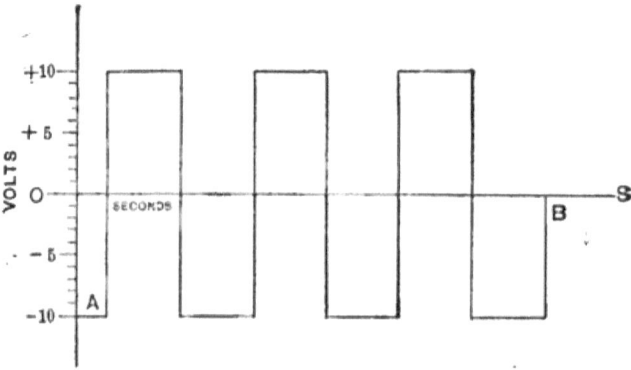

Fig. 43.—Alternating E. M. F.

alternating. Thus, in Fig. 43, the E. M. F. is seen to alternate between 10 volts positive and 10 volts negative, the transitions in this particular case being made instantaneously. Such an E. M. F. might be produced by connecting a battery of voltaic cells with a current reverser, in such a manner, that by rotating the handle, the E. M. F. would be periodically reversed without being withdrawn from the circuit.

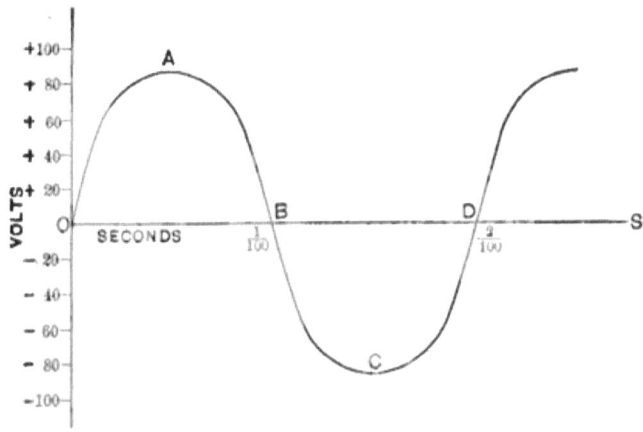

Fig. 44.—Symmetrical Alternating E. M. F.

It is not necessary that an alternating E. M. F. should change abruptly from its maximum positive to its maximum negative value. In most cases, in fact, the change occurs more gradually, as shown in Fig. 44, which represents a common type of alternating E. M. F. Figs. 45 and 46, represent the same alternating E. M. F., although the graphical appearance of the waves is changed, owing to the varia-

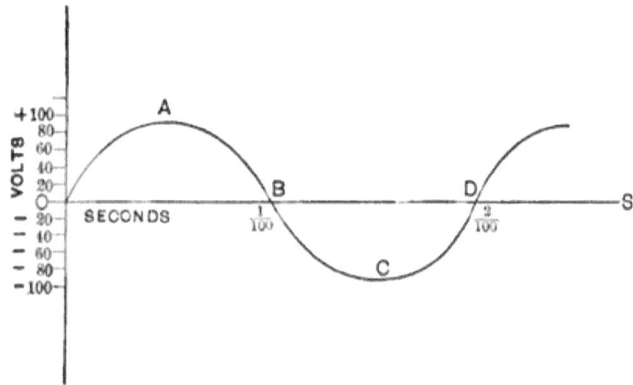

Fig. 45.—Symmetrical Alternating E. M. F.

tions of the scale of time along the base, and the scale of E. M. F. along the vertical. It will be observed that in all representations of alternating E. M. F. there is a motion in one direction, in which the E. M. F., beginning at the base line or zero, gradually increases in value, and then gradually falls until it again reaches zero, then changing its direction and going through the same

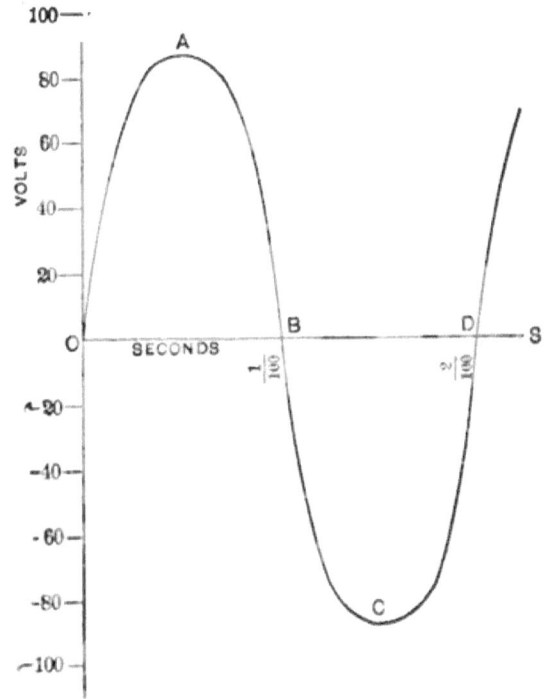

Fig. 46.—Symmetrical Alternating E. M. F.

changes in the opposite direction. Thus the E. M. F. commencing, in Fig. 44, at zero, advances in the positive direction to its maximum or greatest value at A, then diminishing, but still in the positive direction, to zero, at B, changing direction and increasing negatively to a maximum value at C, and then regularly decreasing again to zero where it again repeats the former movement. Each of the waves OAB, or BCD, is called an *alternation;* so that it will be seen, that in the cases considered in Figs. 44, 45, and 46, an alternation lasts 1/100th of a second, or, there are 100 alternations in a second. A complete to-and-fro motion is called a *cycle*, and the time required to complete a cycle is called a *period*. The period in the cases here considered is 1/50th second and it is evident that the number of cycles in a second will depend upon the value of the period. For example, if the period is 1/100th of a second, there would then be 100 cycles in a second. The number of cycles in a second is called the *frequency;* so that, in the last case the frequency would be 100. This is often written 100~.

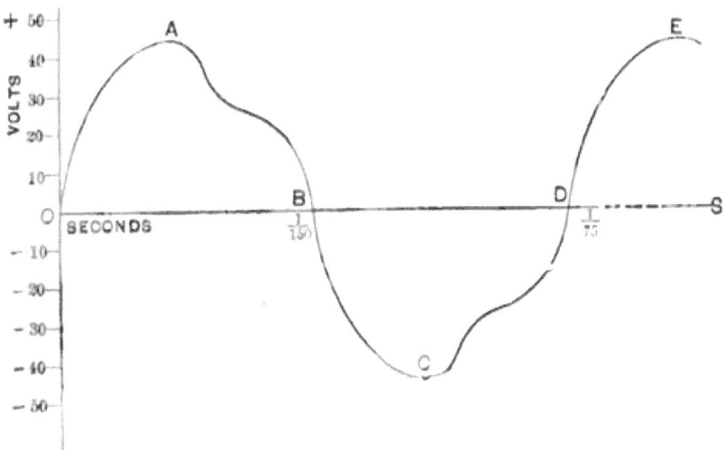

Fig. 47.—Symmetrical Alternating E. M. F.

Alternating E. M. F.'s may be *symmetrical*, or *dissymmetrical*. A symmetrical E. M. F. is one which is graphically symmetrical about its zero line; that is to say, the positive waves are the same as the negative waves, except that they occur in the opposite direction. Fig. 47, represents a type of *symmetrical wave of E. M. F.* having a frequency of 75~; and, consequently, a period of 1/75th second. *A dissymmetrical alternating E. M. F.* is one in

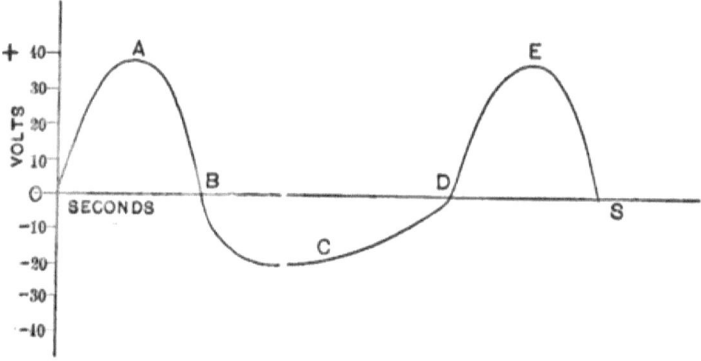

Fig. 48.—Dissymmetrical Alternating E. M. F.

which the positive wave differs from the negative wave not merely in direction but also in outline. A type of dissymmetrical wave is shown in Fig. 48.

Symmetrical alternating E. M. F. waves are produced by *alternating-current dynamos*, or *alternators*. Dissymmetrical alternating E. M. F. waves are produced by particular types of apparatus, such as *faradic coils*, the construction of which will be explained hereafter.

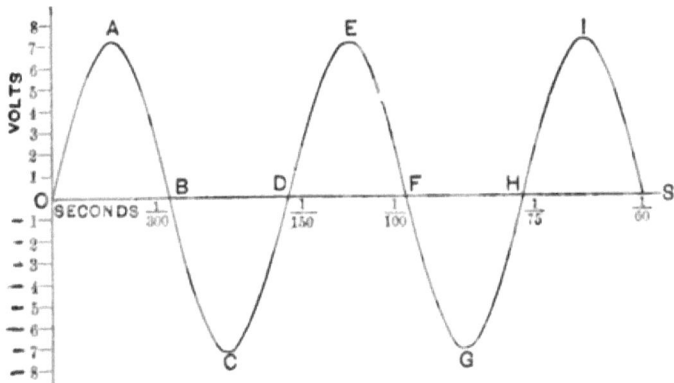

FIG. 49.—SINUSOIDAL E. M. F.

It is evident from the preceding figures that an E. M. F. becomes alternating if it periodically changes its direction and magnitude, and that considerable variation may exist in the manner in which both of these changes may occur. A wave of the form shown in Fig. 49, is called a *sinusoidal wave*, and an E. M. F. alternating in this manner is called a *sinusoidal E. M. F.* This type of E. M. F. possesses important characteristics.

It is evident that the following distinct types of E. M. F. exist; namely,

E. M. F.
- Continuous
  - Pulsating
    - Intermittent.
    - Non-Intermittent.
  - Steady
- Alternating
  - Symmetrical
    - Sinusoidal.
    - Non-sinusoidal.
  - Dissymmetrical

A continuous E. M. F. acting in a closed circuit produces in it a *continuous current*. A pulsating E. M. F. similarly produces a *pulsating current*. An alternating E. M. F. produces an *alternating current*.

In fact, all the varieties of E. M. F. in the above table, when acting in a circuit, produce their particular type of electric current, although the graphic representation of the current is not always the exact counterpart of the graphic representation of the E. M. F. The above table may, therefore, be repeated for currents as well as for E. M. F.'s.

The character of the electric current is dependent on the character of the E. M. F. producing it. A continuous electric current, like the continuous E. M. F. which causes it to flow, does not vary in its strength at different times, but flows through the circuit like a steady flow of water through a pipe or river channel. A continuous electric current is sometimes called a *direct current*, in contradistinction to an alternating electric current, which, like the E. M. F. producing it, changes its direction at every half cycle.

All electric currents, however, are produced by the action of some form of E. M. F., so that the presence of current in any circuit necessitates the existence of an E. M. F. which has produced it.

CHAPTER VI.

ELECTRIC WORK AND ACTIVITY.

An electric current is never produced without an expenditure of work. The greater the strength of the current, the greater is the amount of work done in a given time. In mechanics, the amount of work done by the action of any force is measured by the distance through which the force acts. A convenient *unit of work* is the *foot-pound*, or the amount of work done when one pound is raised, against the force of gravity, through a vertical distance of one foot. Since to do work requires the expenditure of energy, the rate at which work is done represents the rate at which energy is expended. Thus, a pound weight raised through a foot, requires, as we have seen, the expenditure of a foot-pound of work. The same amount of work would be done if a pound weight were raised through a foot in a minute, or in a second, but the rate at which energy would be expended would be sixty times greater in the second case than in the first.

The *rate of doing work*, or expending energy, is called *activity*. For the *electric unit of work* the foot-pound might be employed, but it is more convenient in practice to employ a unit called the *joule*. The joule is nearly equal to 0.738 foot-pound; that is to say, one foot-pound is an expenditure of work equal to about 1.355 joules. In the same way the foot-pound-per-second, the mechanical unit of activity, might be employed as the electric unit of activity, but it is more convenient to employ the *joule-per-second*, or the *watt*, of which 746 are equal to one horse-power, of 550 foot-pounds per second.

An electromotive force or pressure is always required to send a current through a circuit; that is to cause electricity to flow, and, as in the case of mechanical work, the amount of electrical work done is equal to the amount of electricity set in motion multiplied by the pressure which causes it to move. When a unit of E. M. F., or one volt, causes one coulomb of electricity to pass through a circuit, there is one unit of work expended called the *volt-coulomb*, or the *joule*. When a pressure of one volt causes a coulomb-per-second to pass through the circuit, that is when a volt causes a current of one ampere to flow, electric work is done in the circuit at the rate of one watt, so that if we multiply the volts in the circuit by the amperes, we have the activity in the circuit expressed in watts. Thus, if an electric pressure of 20 volts, measured between electrodes, be applied to the human body, and the current produced in the circuit under these conditions be 25 milliamperes, then the electric activity in the body will be $20 \times \dfrac{25}{1,000} = \dfrac{500}{1,000} = 0.5$ watt, or half a joule each second; *i. e.*, 0.369 foot-pound each second.

If a galvano-cautery be supplied with a current of 20 amperes, under a pressure of 2 volts, the activity in the knife will be $2 \times 20 = 40$ watts $= 40$ joules-per-second $= 29.52$ foot-pounds-per-second; and, if this activity be sustained for two minutes, the work done will amount to $40 \times 120 = 4,800$ joules $= 3,546$ foot-pounds; that is to say to 1 pound lifted 3,546 feet, or to one ton of 2,000 pounds lifted 1.723 feet.

The amount of activity expended in a small incandescent lamp, of say 1/2 candle power, frequently employed for exploring cavities in the human body, may be found from the fact that such a lamp only requires a current of 1.4 amperes, at a pressure of 3 volts at its terminals. This represents an activity of $3 \times 1.4 = 4.2$ watts, or an expenditure of 8.4 watts per candle.

Electric activity in a circuit has always to be provided by the source of the driving E. M. F. Thus, when a voltaic battery is supplying a pressure which drives a current, it is the chemical energy in the battery which has to provide the activity and the work done. In other cases, where a dynamo is the source of E. M. F., the power has to be supplied from the engine which drives the dynamo. When, therefore, an E. M. F. drives a current, it does work on that current, and the work must be supplied by the source of E. M. F. On the other hand, when an E. M. F. is driven by a current, that is to say when a current passes in a circuit against the action of an E. M. F., so that the current overcomes the E. M. F., then work is done *upon* the E. M. F. instead of *by* the E. M. F., and work appears in the source of E. M. F. For example, when a current passes through a *voltameter; i. e.*, a vessel containing acidulated water, and provided with platinum electrodes, an E. M. F. is set up at the surface of the immersed electrodes, opposing the passage of the current. Such an E. M. F. is called a *counter E. M. F.*, and is abbreviated C. E. M. F.

Fig. 50, represents a form of voltameter. The current passes between the binding posts through the acidulated water contained in the vessel *A*, being led in and out by platinum plates or electrodes. If a current of two amperes passes through such an apparatus, and a C. E. M. F. of 2.5 volts be set up at the surface of the electrodes, an activity will be expended of $2.5 \times 2 = 5$ watts, upon this E. M. F., and this work will be expended chemically, in decomposing the acidulated water and liberating its constituent gases, oxygen and hydrogen, which appear at the terminals connected respectively with the positive and negative poles of the battery. Assuming that none of the liberated gas is dissolved, it will accumulate in the

ELECTRO-THERAPEUTICS. 131

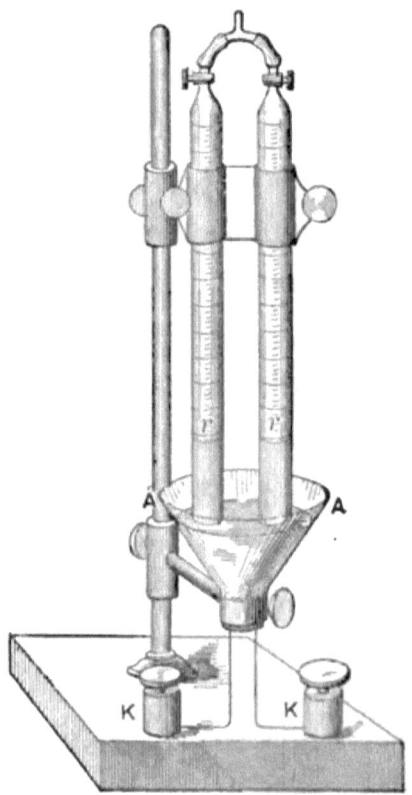

Fig. 50.

collection tubes over the respective electrodes, and, if allowed to enter into combination at any subsequent period, will liberate an amount of work in the explosion, equal to the work done by the electric current in evolving it. A voltameter is used for measuring the strength of a current by the rate of decomposition of water. It is not, however, as convenient for such purposes as an ammeter.

When an electric current passes through a wire offering a resistance, a C. E. M. F. is practically developed in the wire; that is to say, if a current of 5 amperes passes through a resistance of 2 ohms, a C. E. M. F. of 10 volts will be established in the wire, for the reason that, by Ohm's law, 10 volts are necessary at the terminals of the wire in order to send five amperes through 2 ohms resistance. The product of the C. E. M. F. and the current, represents the activity expended on the C. E. M. F., and this work is always expended in heating the wire. In the case assumed, 50 watts would be expended in the substance of the wire as heat. The apparent C. E. M. F., which is produced in a circuit by its resistance when overcome by a current, expends activity in heat; whereas, the C. E. M. F., which is due to chemical decomposition, or to magnetic action, expends activity chemically or magnetically. In other words, work done in a circuit against the C. E. M. F. of resistance, is work expended thermally; and, therefore, is practically irrecoverable, while work done in a circuit against the C. E. M. F. of chemical decompositions, or of magnetic action, is capable of being partly or almost entirely utilized.

Fig. 51, represents an electric *calorimeter*; i. e., a device for measuring an electric current by the amount of heat produced in a wire which is immersed in a known volume of water. In the form of calorimeter shown in the figure, the current enters and passes through the resistance coil NM, surrounded by a known quantity of water. A thermometer T, is

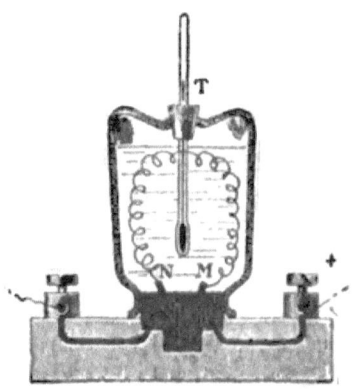

Fig. 51.

provided for measuring the increase in temperature. The amount of electric energy, which must be expended as heat in order to raise the temperature of one cubic centimetre, or one gramme, of water through 1° C., is called a *water-gramme-*

degree-centigrade, or a *lesser calorie*, and is approximately equal to 4.2 joules. Consequently, a pound of water being 453.6 grammes, when raised through a temperature of 10° C., requires an expenditure of energy equal to 4,536 water-gramme-degrees, and $4{,}536 \times 4.2 = 19{,}051$ joules.

In the case of a current of 25 milliamperes passing through the human body, under a pressure of 20 volts, and representing an activity of 0.5 joule-per-second, or 0.5 watt, a certain amount of this C. E. M. F., probably 2.5 volts, would be due to electrolytic action, and the remainder, or 17.5 volts, due to the resistance of the body as a conductor. The activity expended electrolytically would, therefore, be $2.5 \times \dfrac{25}{1000} = 0.0625$ joule-per-second, and the remainder, or $\dfrac{17.5 \times 25}{1{,}000} = 0.4375$ joule-per-second, would be expended in warming the conducting materials in the body. The electrolytic work would be expended at the surface of the metallic electrodes, while the thermal activity would be expended wherever the resistance was overcome; and, moreover, expended in proportion to the amount of resistance. Assuming, however, for the sake of illustration, that the amount of heat so developed was equally diffused throughout the whole body, and that the capacity of the body for heat was that of 100 lbs. of water, then one joule, expended in the body thermally, would raise its temperature $\frac{1}{4.2 \times 100 \times 453.6} = \frac{1}{190,510}$ of a degree centigrade. 0.4375 joule would raise it $\frac{0.4375}{190,510}$ of a degree centigrade. The application of the current for 30 minutes, or 1,800 seconds, would raise its temperature $\frac{787.5}{190,510} = \frac{1}{242}$ of a degree centigrade, approximately.

It will therefore be evident that the current strengths ordinarily employed in electro-therapeutics cannot directly raise the temperature of the body to any sensible degree, although physiological activities called into action thereby may do so. A current of several amperes, however, passed through the body, for a few seconds, will raise the temperature appreciably.

CHAPTER VII.

FRICTIONAL AND INFLUENCE MACHINES.

OUR earliest notions concerning electricity were obtained from the electric effects produced by the friction of one substance against another, such, for example, as a piece of glass against a silk handkerchief. Under these circumstances, both the glass that is rubbed and the silk with which it is rubbed, acquire electric excitement, as manifested by their ability to attract light bodies, such as shreds of paper, brought near them. It can be shown that all bodies possess the ability of acquiring electric excitement by friction against other bodies, and that whenever two dissimilar substances, or even two dissimilar surfaces of the same substance, are brought into contact, an E. M. F. is produced at the contact surfaces, one substance becoming positive and the other negative. The friction of one substance against another is, therefore, another method of producing an E. M. F., and, as in the case of the other sources referred to, this E. M. F., if provided with a circuit, is capable of setting electricity in motion.

The E. M. F.'s produced by friction, are much higher than those produced by either voltaic cells or dynamo-electric machines; consequently, they are capable of causing electricity to pass through a circuit even when separated by a small interval or *air-gap*. It is found that an E. M. F. of, approximately, 80,000 volts is required to produce a discharge or spark across an air-gap one inch in length, between two slightly convex surfaces, and that, roughly, 80,000 volts per inch of sparking distance, is a fairly reliable estimate of pressure for distances, lying between 1/100″ and 3″. Beyond these limits, the rule cannot be safely applied, and, in fact, for very large sparking distances of more than one foot, a much smaller E. M. F. than 80,000 volts per inch appears to be needed. When the opposed electrodes terminate in points, instead of in blunt surfaces, a relatively much smaller pressure is needed to effect discharge.

Various devices have been employed in order to produce electricity by the friction of one substance against another. Such devices are called *frictional electric machines*, and consist essentially of a plate or cylinder, generally of glass, so rotated as to be rubbed against a *rubber* of chamois skin, covered with an *amalgam* of mercury and tin. By this friction, both the rubber and the glass acquire an electric potential. A *comb of points* placed near the glass and connected with an insulated conductor, enables the conductor to become charged. A smaller insulated conductor, connected electrically with the rubber, enables the conductor to take the opposite or negative charge.

A well known form of frictional electric machine is shown in Fig. 52. Here the glass plate PP, is rotated in a vertical plane between two rubbers R, R, of chamois leather, connected electrically with the ground. In this form of machine, two insulated conductors C, C, are connected to separate pairs of combs near the surface of the revolving glass plate and thus receive a positive charge.

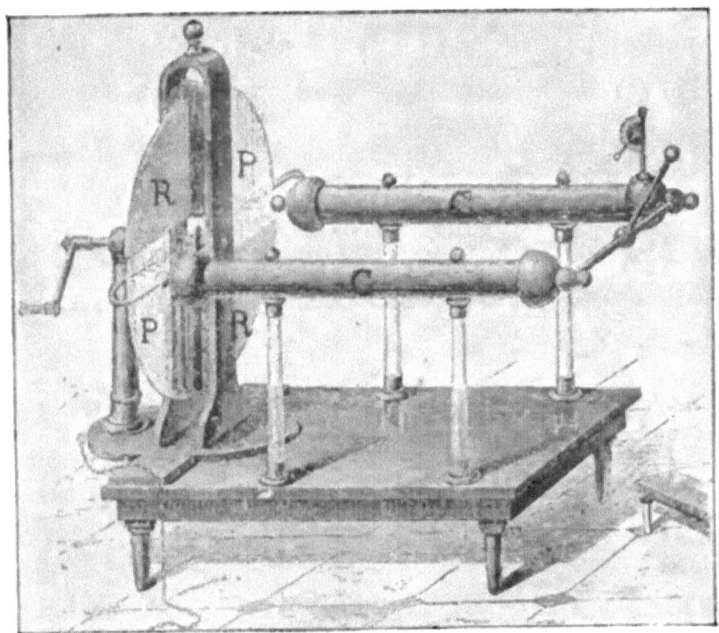

Fig. 52.

When an electric machine is properly operated it has the power of sending a torrent of minute sparks through a considerable air-space. The E. M. F. produced by this type of source is of the pulsatory character, and is shown in Fig. 53, here represented at about 140,000 volts, or 140 kilovolts.

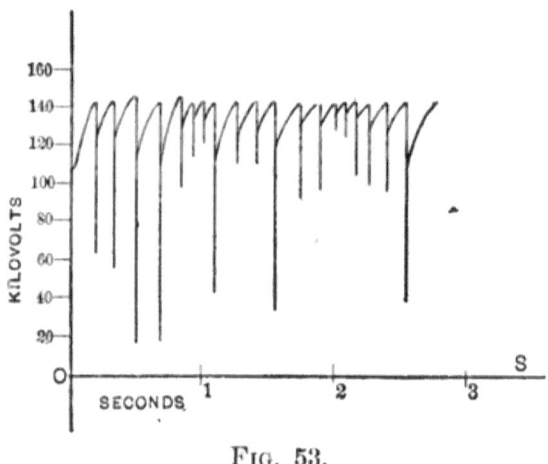

Fig. 53.

When a pulsatory E. M. F. rises to such an amount as will permit it to discharge through an air-gap, it suddenly falls on discharge to a minimum which is not always the same. It then recovers and again discharges, this action being carried on in a pulsatory manner at frequent intervals. E. M. F.'s of this character are frequently employed in electro-therapeutics under the name of *Franklinic E. M. F.'s*, after Benjamin Franklin. A frictional electric machine, however, is too uncertain in its action to be employed for this purpose, being too much dependent on the conditions of the weather, since, during damp weather, the film of moisture which settles upon the surface of the glass, and on the supporting pillars, is often sufficient to conduct away the electric charges and prevent their formation. For this reason, frictional machines have been replaced by *influence machines* of which there are a number of different designs.

In order to understand the operation of an influence machine, it is necessary to first investigate what occurs in the space surrounding an electrically charged body. If we support a metallic sphere *A*, upon a table in a room, *B C D E*, Fig. 54,

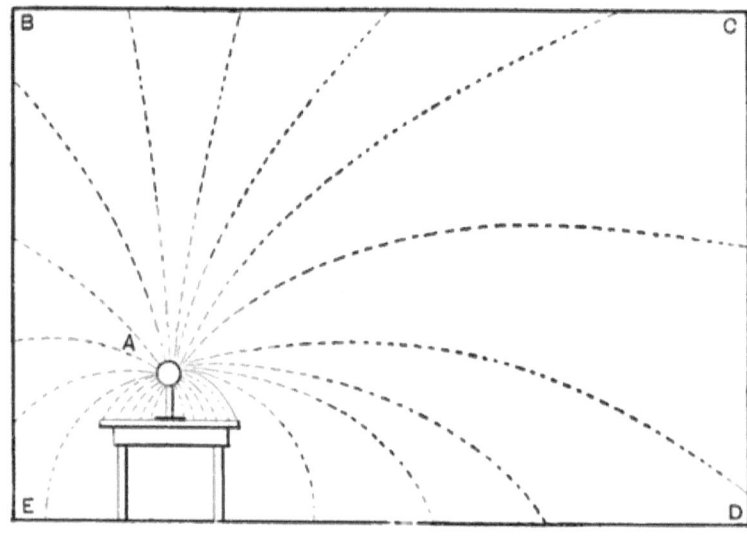

Fig. 54.

and connect this sphere with an E. M. F., the sphere will receive a charge. If we connect the sphere by a wire to one terminal of a voltaic cell, the sphere will become charged, thereby, but so feebly, that delicate apparatus will be necessary in order to reveal the presence of the charge. If, however, we connect the sphere to one terminal of a battery of 10,000 cells, each having an E. M. F. of one volt, an appreciable charge will be communicated to the sphere, which will now manifest distinct electrostatic properties. Finally, if we connect the sphere to one terminal of an electrostatic machine, having an E. M. F. of, perhaps, 200,000 volts, the charge acquired by the sphere will be comparatively great. That is to say, it will receive a comparatively large quantity of electricity, which will be a certain fraction of a coulomb. It is common to regard such an electric charge as being situated on the surface of the body A. Such, however, is not the fact, the charge really resides in the air, or more strictly in the air and the ether surrounding the body, and the function of the metallic cylinder is merely to provide a surface from which the charge can enter and influence the ether around it.

If we assume, for the sake of illustration, that the charge communicated to the sphere is $\frac{1}{1,000,000}$th of a coulomb, or one *micro-coulomb*, and that the E. M. F. at which it was delivered to A, was 100,000 volts, or, in other words, that the potential of A, is 100,000 volts, then the work delivered to A, is $100,000 \times \frac{1}{1,000,000} = \frac{1}{10}$ joule=0.0738 foot-pound. This energy is received by the air and ether surrounding the sphere, and is held there during the maintenance of the charge. The energy is distributed through all the ether in the room, although not equally distributed.

A certain fraction of a joule is thereby charged in each cubic inch of space, the greater amount being in the immediate neighborhood of the sphere, and lessening with distance from the same. The charge is passed into the ether by an action which is called *electric displacement*. Electric displacement takes place along defined lines or curves through the ether in the room, or as it is sometimes stated, *electrostatic flux* proceeds from the charged body through the ether in the room along lines or paths called *lines or curves of electrostatic flux*. The shape of these lines depends on the shape of the body, and on the shape of the enclosure in which it is placed, or of the bodies which may be in its neighborhood. Fig. 54, gives a diagrammatic view of some of these displacement curves, or curves of electrostatic flux. The total amount of electrostatic flux is proportional to the charge on the body.

Fig. 55, gives a graphic representation of a mechanical model showing the action

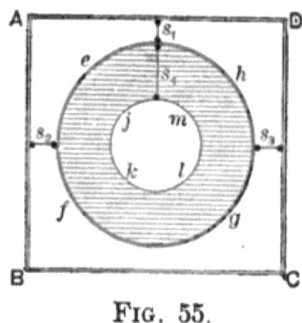

FIG. 55.

which an electrified sphere exerts upon the space surrounding it. Let $j\ k\ l\ m$, be a diametral section of a spherical elastic bag of rubber, and $e\ f\ g\ h$, the section of another bag of rubber surrounding the first concentrically. Moreover, suppose these bags to have the space between them entirely filled with water. If now, air be pumped into the interior of the inner bag it will distend, and, in distending, will cause the outer bag to also distend, although to a lesser degree, since the water between them is practically incompressible. The inner bag corresponds to the metallic sphere of Fig. 54, and the outer bag corresponds to the walls of the room, in which the sphere A, is suspended. The water filling the space between the bags corresponds to the ether filling the space between the sphere and the walls of the room. The air pressure communicated to the interior bag by pumping air into it, corresponds to the electric pressure, or high potential communicated to the sphere. Under the influence of this pressure, the inner bag expands or receives a charge, the expansion causes liquid flux or streamings through all the mass of liquid, which distends the outer sphere. This corresponds to the fact that the charge communicated to the sphere is imparted to the surrounding ether and passes, in the form of electrostatic flux, through the whole ether space until intercepted by the walls of the room. The total charge of the sphere may be regarded as being equal to the total displacement and also numerically equal to the total negative charge accumulated on the walls of the room.

In order to show the effect of the shape of a body on the direction of the flux paths, a few examples may be noted. Fig. 56, shows two insulated concentric spheres connected with an E. M. F., the inner sphere being positive. Here the flux issues from the inner sphere in radial lines, and terminates on the surface of the outer sphere. The total amount of displacement which passes through any spherical envelope, which could be drawn between A and B, is equal to the charge which resides on A or B. In other words, the system behaves as though a certain amount of ether had been liberated at the surface of A, passed through the surround-

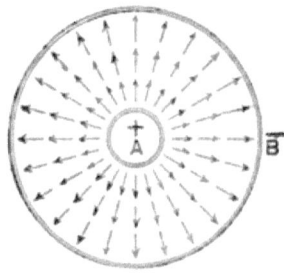

Fig. 56.—Electrostatic Flux Paths Between Concentric Spherical Conducting Surfaces, Insulated from Each Other.

ing space against elastic reaction, along radial lines, until finally, a certain amount is forced into the shell of the external sphere at B. The quantity of charge which thus enters the system depends upon two things; namely, the magnitude of the E. M. F. employed, and, secondly, the shape of the mass of ether brought under the action of this force. If we double the E. M. F., we double the charge, and if we halve the E. M. F. we halve the charge. Again, if we employ a

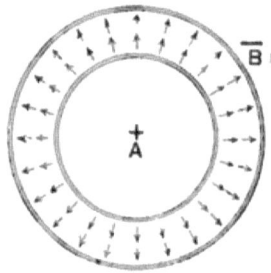

Fig. 57.—Electrostatic Flux Paths Between Insulated Concentric Spheres.

thinner stratum of ether we shall reduce its resiliency, and increase the charge for a given E. M. F. employed. If, for example, the inner sphere be larger, as shown in Fig. 57, the same E. M. F. will now act upon a thinner layer of air and ether, and will produce a greater displacement or

154 ELECTRICITY IN

charge; in other words, the resiliency of the mass of ether, enclosed between the ter-

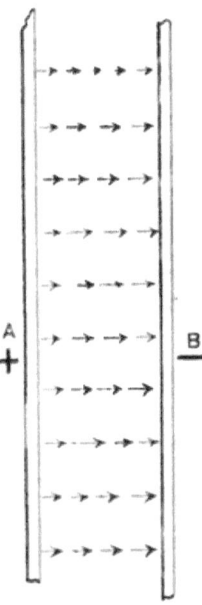

Fig. 58.—Electrostatic Flux Paths, Parallel Plane Spheres.

minals connected with the E. M. F., has been reduced.

Fig. 58, represents two parallel plane surfaces connected with an E. M. F.

Here the left-hand plane A, is considered as positive; *i. e.*, the electrostatic flux is assumed to emanate from A, pass through the intervening space, and terminate at the surface of B. There will be a positive charge on A, an equal negative charge on B, and the same charge represented in displacement all through the mass of ether between the plates. The amount of charge which will enter the system, will depend upon the E. M. F. brought to bear upon the plates A and B, the thickness of the stratum of ether, and the area of the plates or stratum. If the area of the opposed plates be increased, the elastic resiliency of the mass of ether between the plates is diminished, and a proportionally greater charge enters the system. Similarly, if the plates be approached, so that the stratum of included air becomes thinner, its resiliency is diminished, and the E. M. F. will force more flux through the system, and a corresponding greater charge. In either case, therefore, we have what might be regarded as an *electrostatic circuit.*

The amount of flux which will pass through the circuit; *i. e.*, the amount of charge which can be communicated to the surface of the dielectric involved, will depend upon the E. M. F., and also upon the elastic resistance of the medium.

$$\text{Electrostatic flux} = \frac{\text{E. M. F.}}{\text{Electrostatic resistance.}}$$

The greater the electrostatic resistance, the less the flux, and vice versa. This corresponds completely to Ohm's law for the voltaic circuit, except that the *electrostatic resistance* is not a resistance to the passage of electric current but is the resistance to the passage of *electrostatic current* or flux. Moreover, the same rules apply to the resistance offered by a wire to the passage of a current, and the resistance offered by a dielectric mass to the passage of an electrostatic flux. The longer the mass the greater the resistance; the greater its area of cross-section, the smaller its resistance.

The *displacement lines*, or lines of electrostatic flux, which may be drawn for any completely specified electrostatic system, and which can be experimentally determined in most cases, represent lines in the dielectric medium along which stress exists, by virtue of the electrostatic flux. This stress, which is developed in the ether, is dependent upon the energy absorbed by the ether during the existence of the electric charge. Along these curves, in fact, there is exerted a continual tension, or, in other words, the displacement lines are always tending to contract

and shorten. For example, **the two
charged plates** A **and** B, shown in Fig. 58,
being connected **by a number** of displacement lines, tend to attract each other.
The real tendency is for the shortening of
the lines of stress, or flux lines. The ordinary statement that positively and negatively electrified bodies tend to attract
each other, should more accurately be:
positively and negatively electrified bodies,
being connected by lines of electrostatic
flux, tend to come together by reason of
the contraction of the flux lines.

If, as in Fig. 59, a small spherical conductor C, be introduced into the electrostatic
flux, the effect is twofold; first, the electric
medium is thinned locally by the presence
of the conductor, so that its resiliency is
locally diminished, and a more powerful
flux will pass through the **system** in its

neighborhood than elsewhere. Second, the flux will be intercepted by the conductor, which will form the termination of the flux on the entering side and a new starting point on the leaving side. A negative charge will, therefore, appear on the surface where the flux terminates, and a positive charge on the surface where the flux reissues. This appearance of positive and negative charges, on opposite sides of an insulated body supported in an electrostatic flux, is commonly called *electrostatic induction*. It is merely a consequence of the fact that the body relieves from electrostatic stress the ether which it displaces, and that, in consequence of this relief, charges appear at the surfaces where the flux enters and leaves.

If in Fig. 59, the small conductor C, is charged by having been connected with a suitable E. M. F. it will do more than merely relieve ether of its duties; for, it

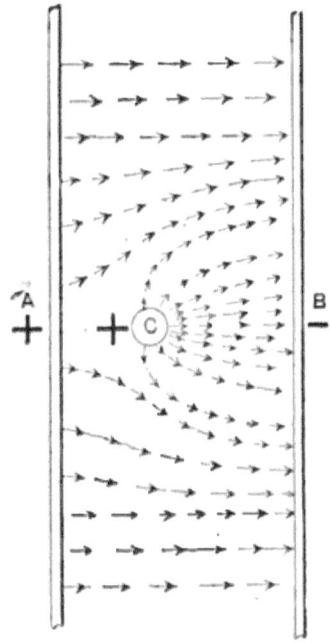

Fig. 59.—Diagrammatic Representation of Electrostatic Flux Paths.

will add flux of its own to the flux in which it is introduced. For example in Fig. 60, two spheres + and −, are shown, at *A*, which have been connected with the positive and negative terminals of a high E. M. F. The electrostatic flux passing between them, through the surrounding ether, is partly represented diagrammatically by the dotted lines. These two spheres evidently behave as though they attracted each other, owing to the contracting forces of all the flux paths between them. If we suppose them fixed upon suitable insulating pillars, so that they cannot approach each other, and that a smaller conducting sphere is introduced between them, as shown, this smaller sphere will acquire a positive charge from the positive sphere, and will thus become the recipient of a number of flux paths which emerge from it, which tend to pull it across toward the negative sphere. Under the influence of these attractive forces, the smaller sphere, if it be free to move, will move to the right. When it is in the position shown on the right, midway between the two spheres, it will be seen that many flux paths connect it with the negative sphere, while no flux paths connect it with the positive sphere. As soon

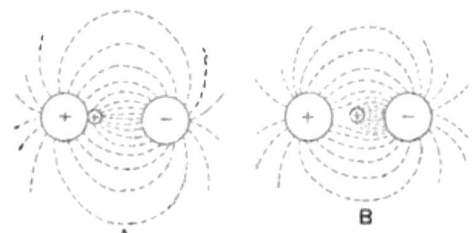

Fig. 60.—Effect of Charged Sphere.

as it reaches the negative sphere it will deliver up its charge, and reduce the potential of the negative sphere unless the latter be connected with an electric source. It will then acquire a negative charge and new flux paths will enter its surface from the positive sphere. Owing to the attraction of these flux lines it will again be drawn to the left and thus a continual to-

and-fro motion will be set up, so long as a
difference of potential or E. M. F. exists
between the two large spheres.

Fig. 61, represents at A, the effect of
inserting, between the two large spheres,

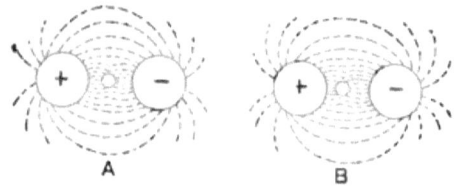

FIG. 61.—EFFECT OF UNCHARGED SPHERE.

a small sphere in an uncharged condition.
It will be observed that the effect is to
intercept a larger number of flux paths,
and thus to relieve from duty the ether
contained within the space occupied by
the small sphere. In this case the attractions, on each side of the small sphere, are
balanced. If, however, the small sphere
be placed nearer one side than the other,

as shown at B, the stratum of ether between it and the positive sphere will be thinner than the stratum on the right; and, consequently, a greater electrostatic flux will pass through the space on the left, thereby entailing the introduction of a greater number of flux lines, and a greater electrostatic force urging the sphere to the left.

It will be seen, from a consideration of the preceding phenomena, that the following may be generalized as the laws of electrostatic charges, attractions and repulsions; namely,

(1) That every electrified body forms a locus or place where electrostatic flux enters or leaves a dielectric medium; and, conversely, that all conducting surfaces, where lines of electrostatic flux terminate, are said to be charged surfaces, the flux

being assumed to leave at positively charged surfaces and to enter at negatively charged surfaces.

(2) Lines of electrostatic flux are the directions along which electrostatic stress exists in the ether, and accompany the temporary absorption of energy into the ether.

(3) Dissimilarly charged bodies attract one another, because lines of electrostatic flux tend to contract.

(4) That similarly charged bodies appear to repel, owing to the fact that no flux paths connect them, but that flux paths connect each of them with neighboring objects, so that they are drawn to the neighboring bodies and are not repelled from each other.

(5) Electrostatic induction accompanies the introduction of a conductor into the electrostatic flux, whereby charges are

caused to form upon opposite sides of the interposed conductor.

The principle of electrostatic induction is employed in influence machines. A

FIG. 62.—ELECTROPHORUS.

simple form of influence machine is called the *electrophoros*, and is illustrated in Fig. 62. A disc B, of vulcanite, resin, or other suitable material, is vigorously rubbed, say with a cat skin, and thereby becomes negatively charged, under the influence of the powerful E. M. F. set up. If such an electrified disc be laid on a table, as shown in Fig. 63, so that its electrified surface is uppermost, the flux paths may be represented diagrammatically by the arrows.

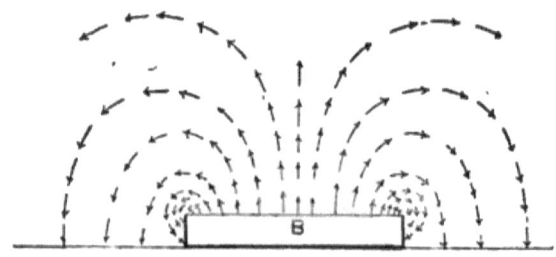

Fig. 63.—Representing Action and Operation of Electrophorus.

If now, an insulated metallic disc *A*, Figs. 62, and 64, furnished with round edges, be rested on the disc, there will be no great change produced in the electrostatic system. There will only be a slight reduction in the *dielectric resistance* of the air, owing to the interposition of the conducting disc across the electrostatic flux paths. When, however, **the** disc is touched with a finger, as shown in **Fig. 62,** or connected with the ground, as shown in Fig. 65, all the flux paths are shortened, until they exist only between

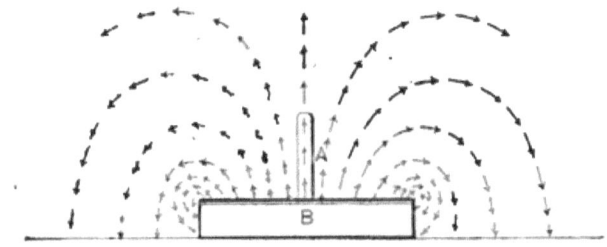

Fig. 64.—Representing Action and Operation of Electrophorus.

the excited disc and the **grounded** metallic plate. In other words, the length of the electrostatic circuit **has** been reduced to that of a thin film of air, and, consequently, the electrostatic **flux,** set up across this film, will be comparatively powerful, and a powerful charge will **be** communicated to the plate at the point where the flux emerges from it. If now, the ground connection to the plate be removed, and the plate lifted, as shown in Figs. 66 and 67, the electrostatic circuit will again be lengthened, and a

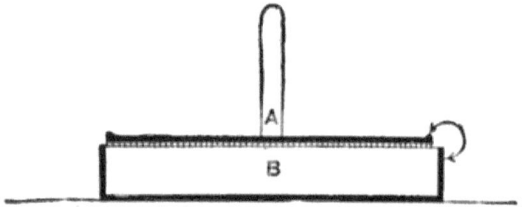

Fig. 65.—Electrophorus.

charge will be left in the plate as well as on the disc, the disc being negative and the plate positive. The metal plate can be discharged and recharged many times in succession.

An *influence machine* is, in reality, a form of a revolving electrophorous. A common form of influence machine, called a *Toepler-Holtz Machine* is shown in Fig. 68. It will be understood that since this apparatus operates by electrostatic induction, no friction is needed. An initial charge is, however, required. The apparatus consists essentially of three

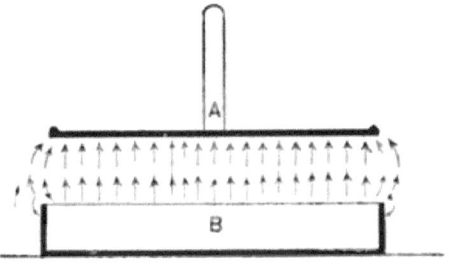

Fig. 66.—Electrophorus.

vertical glass plates of which the central is the largest in diameter, and is fixed, while the two outside plates are mounted on a common shaft, and are capable of being revolved in the same direction by the aid of the handle *h*. The central plate carries two metallic and papered surfaces

ELECTRO-THERAPEUTICS. 171

$A B C$ and $A' B' C'$, one of which $A B C$, is positively electrified at the outset and $A' B' C'$, is negatively electrified. The outside plates carry only metallic buttons on their external surfaces, each button

FIG. 67.—ELECTROPHORUS.

consisting of a disc of tinfoil, with a small brass cap in the centre. Six of these tinfoil discs are represented as being carried on each outside plate.

An electrostatic circuit will be set up from the positively electrified to the negatively electrified surface as shown at *A*, in Fig. 69. The presence, however, of the metallic rod *RR*, which is supported in

Fig. 68.—Triple-Plate Toepler-Holtz Electrical Machine.

such a manner that the combs at its extremities come in contact with the insulated discs on the outside plates as they revolve, limits the electrostatic circuit almost entirely to the space between the central and outside plates, as shown at *B*, Fig. 69, so that the flux paths become more numerous and terminate on the inner

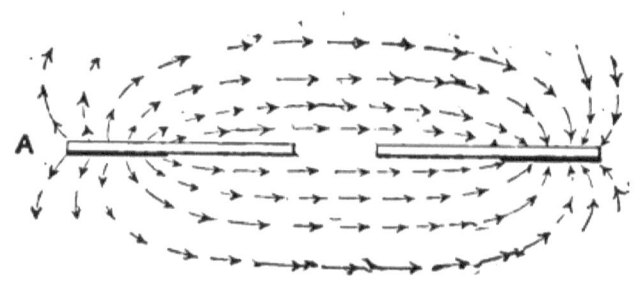

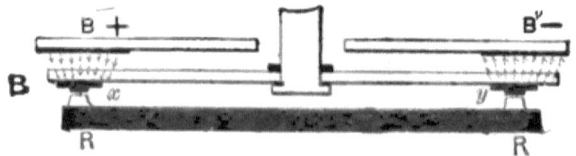

Fig. 69.—Electrostatic Circuits of Influence Machine.

surfaces of the insulated discs as they pass by. Under these circumstances, a negative charge will form on the disc *x*, and a positive charge on the disc *y*. As soon as the discs have been carried from beneath the combs on the rod RR, they retain these charges until they reach the opposite side of the frame, when the disc x, comes in contact with the brush b', thereby communicating its charge to the already negatively electrified surface $A'\ B'\ C'$, on the central plate, and, passing with the remainder of its charge, delivers this remainder to the comb of points attached to the handle and main conductor H'. Similarly, the disc y, which retains its positive charge after quitting the comb on the lower extremity of the rod RR, is carried to the brush b, and communicates its charge to the already positively electrified surface $A\ B\ C$, the remainder of its charge being collected by the comb on the handle H. Consequently, during rotation, the half of the rotating plates on one side of the rod RR, is positively, and the other half, negatively electrified. The charges on the electrified surfaces $A\ B\ C$ and $A'\ B'\ C'$, automatically increase, until a balance is maintained between the further accession of charge, and the leakage which takes place between them. This leakage limits, therefore, the maximum E. M. F. obtainable by the machine. When the discharging rods *II, II*, are brought close together, the pressure obtained is lower, owing to the fact that a smaller E. M. F. is required to produce a spark discharge across the air-gap, and a more rapid stream of discharges over this air-gap and a lower pressure may, consequently, be expected. On increasing the distance between the discharging rods, the pressure increases, but the frequency of discharge usually diminishes.

A form of apparatus known as a *condenser* consists essentially of an electrostatic circuit of low resistance, that is to say, of an electrostatic circuit of short length and large cross-sectional area. A condenser, therefore, offers a comparatively small elastic resistance to displacement of flux, and, under a given E. M. F., will receive a correspondingly large charge.

Fig. 70, shows a Leyden jar, which is the usual form of condenser employed with high E. M. F.'s. Here the active surfaces are formed of inner and outer coatings of tin-foil, and the dielectric consists of the glass walls of the jar. The length of such an electrostatic circuit; *i. e.*, the thickness of the glass, may be about 1/8th of an inch, and the cross-sectional area of the electrostatic circuit; *i. e.*, the area of the tin-foil surface, about a square foot. Moreover, glass offers less electrostatic resistance than air, and, therefore, the glass

Leyden jar makes a better condenser than an imaginary air jar of the same dimensions. The relative value of the glass de-

FIG. 70.—LEYDEN JAR.

pends upon its quality, but it may readily offer five times less electrostatic resistance than air; consequently, the capacity of a

Leyden jar condenser may be five times greater than that of a similar air condenser. Two small Leyden jars are shown in Fig. 68, having their inner coatings connected with the main terminals H and H', and their outer coatings connected by a metallic strip not shown in the figure. The effect of these jars is to diminish the electrostatic resistance between the terminals, and, therefore, enables a given E. M. F. to accumulate a greater electrostatic flux or charge between the terminals.

The electric energy obtained from the discharge of an influence machine through an external circuit is supplied, mechanically, in the effort necessary to revolve the machine against electrostatic forces. One electrostatic machine acting as a generator, may readily be made to cause another electrostatic machine to run back-

ELECTRO-THERAPEUTICS.

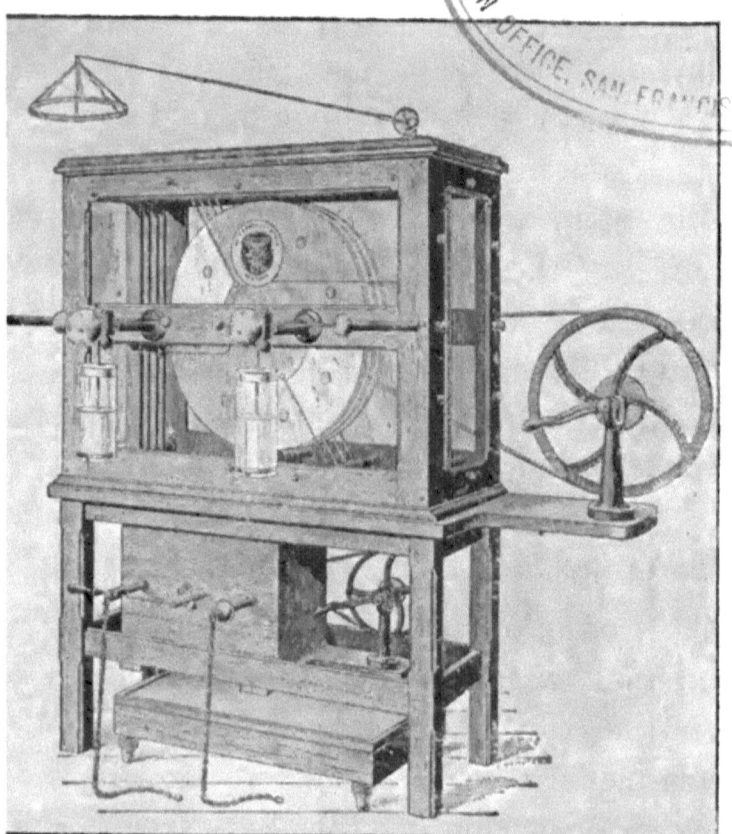

Fig. 71.—Holtz Influence Machine.

wards, as a motor. The hand has, therefore, to be applied with greater force to drive the influence machine, owing to the fact that it is operating so as to furnish current to the circuit connected to it.

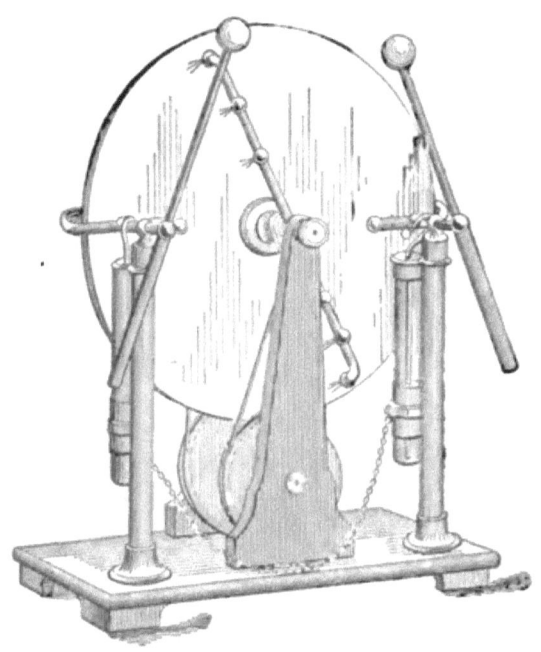

Fig. 72.—Bonetti Electrostatic Influence Machine.

A number of forms of influence machines are in existence. The principal difficulty in operating such machines is to maintain their insulation during all condi-

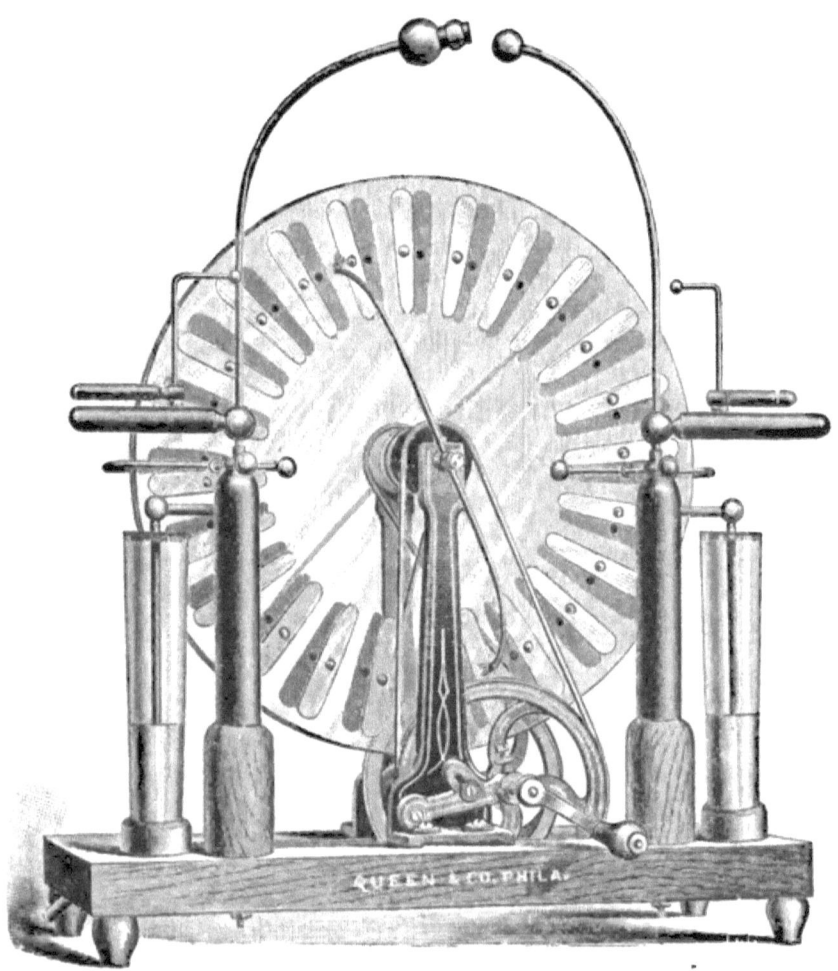

Fig. 73.—Wimshurst Electrical Machine.

tions of weather, so that their charge shall not be lost. For this purpose they are often enclosed in glass chambers in which the air is kept dry by some hygroscopic substance, such as calcium chloride. Such a form of machine driven by a small electric motor is shown at Fig. 71.

In some forms of influence machines, in order to ensure the presence of a small charge, a small frictional attachment is supplied, so that the proper charge shall be ensured by the friction set up.

Glass plates are not invariably used in these machines. Sometimes plates of hard rubber are employed as shown in Fig. 72.

Another convenient form of influence machine is shown in Fig. 73 called the

Wimshurst machine. In this case two glass plates, supporting a number of small, separate tin-foil conductors, are rapidly driven in opposite directions. The action of the machine differs only in detail from that already described.

In conclusion, we may observe that all electrostatic influence machines depend for their operation upon the principle of the electrophorus. The electrostatic circuit in such machines is periodically lengthened and shortened, and the charges so induced are separated and accumulated.

CHAPTER VIII.

MAGNETISM.

MAGNETISM is the science which treats of the properties and laws of magnets whether artificial or natural. Although the nature of magnetism is not known, yet a certain relationship unquestionably exists between magnetism and electricity, so that a knowledge of the nature of one must inevitably determine a knowledge of the nature of the other. Both are believed to be active conditions of the universal ether and are so related that the following laws appear to hold generally; viz.,

(1) A motion of electricity invariably produces magnetism.

(2) A motion of magnetism invariably produces an E. M. F.

The nature of the action which exists between electricity and magnetism may

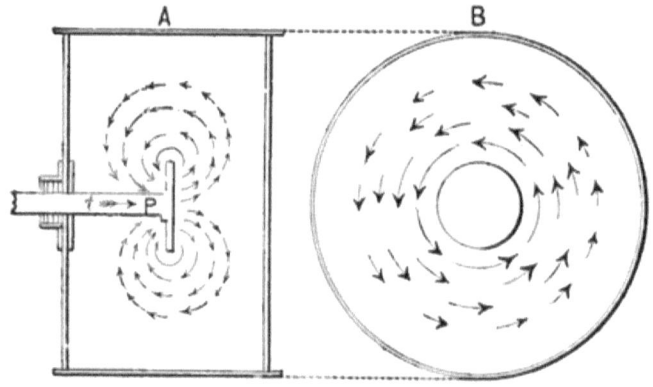

Fig. 74.—Hydraulic Analogy of Relation Between Electricity and Magnetism.

be illustrated by the following hydraulic experiment. Suppose that a large cylindrical tank, represented in Fig. 74, be completely filled with water, and that a plunger P, is provided with a rod t, passing through water-tight packing in the centre of one circular end. It is evident that if the rod t, be moved forward, the plunger P, will advance into the tank. In so doing it will displace the water in front of it, which will flow round to the back of the plunger in vortical paths, formed symmetrically around the face of the plunger. These vortical paths, passing from the front to the back of the plunger, are illustrated diagrammatically at A, Fig. 74. The vortical movement of the water will clearly be most marked in the immediate neighborhood of the edges of the advancing plunger, gradually decreasing from the edges to the sides of the tank. Imaginary lines in the mass of the water, drawn so as to represent the intensity of the vortical movement, form circles, concentric to the axis of the plunger, and at right angles to the direction of its motion, as represented at *B*, Fig. 74. Circles are marked with long arrows near the edge of the plunger where the vortical motion is most intense, and with shorter and shorter arrows at greater distances from it. If now, we remove the plunger from the tank, and artificially cause a system of electric currents to be produced in the mass of quiescent water, such as is represented at *B*, in Fig. 74, then accompanying this system of electric currents, would be produced a magnetic distribution throughout the water, such as is represented by the stream lines at *A*, at right angles to the direction of the electric current. In other words, the relation of magnetic distribution, to electric current distribution, in any space, is identical with the relation between the stream lines of motion in a liquid, and the vortical distribution of motion accompanying the same.

It follows from the preceding that if the all-pervading ether were a non-compressible fluid, like water, and if electric currents consisted of vortices or whirls in this fluid, that magnetism would consist of a streaming motion in the ether. The properties of the ether are not yet thoroughly known, and it is by no means certain that electric currents are vortices therein. All that can be safely asserted is the existence of a relationship between electric activity and magnetic activity in the universal ether, of the general nature we have here pointed out; so that, if we should at any time, discover the nature of either electricity or magnetism, the nature of the other would be immediately deduced.

Magnetism may be produced in two ways; viz.,

(1) By permanent magnets of iron or

steel; or, in a lesser degree, by other magnetic metals such as nickel or cobalt; and,

(2) By electric currents.

Magnetism appears to be an inherent property of the molecules, or ultimate particles, of iron or steel. In other words, if we could isolate and perceive a single molecule of iron, it is believed that we should find that it naturally and permanently possessed magnetism, as a property inherent in it. If the ultimate particles of iron are essentially magnetic, the question naturally arises, why all iron does not manifest magnetic properties? The reason is believed to be found in the fact that in iron, which is apparently unmagnetized, the molecules lack a definite arrangement of direction, and, pointing irregularly, mask or neutralize each other's magnetic influence. Such an undirected system would, therefore, necessarily possess no appreciable external magnetism. When a bar of iron is magnetized, it is subjected to a process whereby its molecular magnets are aligned, or similarly directed, and, acting in concert, are thereby enabled to manifest external magnetic effects.

The region surrounding a magnet is filled with what is called *magnetic flux* or *magnetism*, which is most powerful in the immediate neighborhood of the poles. If we assume, as a working hypothesis, that magnetism consists of a streaming motion of the ether, in accordance with the hydraulic analogue of Fig. 74, then we may regard a magnet as a device for producing such a streaming motion in the ether. The magnetic flux; *i. e.*, the streaming ether, would issue from one pole of the magnet, and, after having passed in expanding curved paths through the space surrounding the magnet, would re-enter it at the other pole. The pole from which the magnetic flux is conventionally assumed to issue is called the *north pole;* i. e., the pole of the magnet, which, if the magnet were freely suspended, would tend to point toward the geographical north; and the pole at which the flux enters the magnet, after having passed through the region outside it, is called the *south pole.* The flux, after entering the magnet, passes through the body of the magnet to the north pole, where it again emerges. Magnetic flux, therefore, completes a closed path or circuit called a *magnetic circuit,* and the convention employed as to its direction in this circuit, is similar to the convention employed as to the direction of electrostatic flux in an electrostatic circuit, or to the direction of a current in a voltaic circuit.

When an electric current passes through a conductor, the conductor temporarily

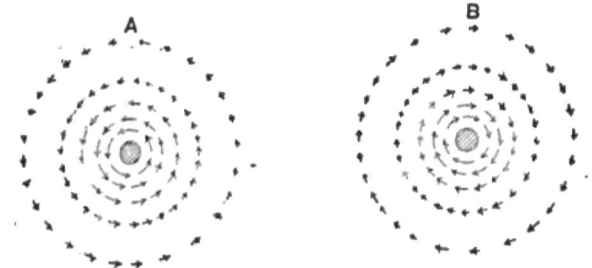

FIG. 75.—MAGNETIC FLUX PATHS SURROUNDING A STRAIGHT ACTIVE CONDUCTOR.

acquires magnetic properties, magnetic flux encircling the conductor in concentric paths. The direction of magnetic flux around an active conductor, depends on the direction of the current in the conductor. This is shown in Fig. 75, where, at B, the current is supposed to be passing through the wire, in a direction from the observer. Here the circles surrounding the wire show that the magnetic flux is passing in concentric circles in the direction of motion of the hands of a clock; while at A, where the current passes through the wire in a direction towards the observer, the direction of the magnetic flux around the wire is opposite to the direction of motion of the hands of a clock, or counter-clockwise. A suspended magnetic needle, introduced into the neighborhood of the active wire; *i. e.*, into the influence of its circular magnetic flux, is deflected thereby, and tends to set itself parallel to the magnetic flux, or at right angles to the direction of the current, its north pole pointing in the direction of motion of the flux.

The power possessed by an active conductor of deflecting a magnetic needle is utilized in a number of ammeters, in which a magnetic needle is deflected by the passage of a current through a number of turns of wire placed in its vicinity. When a wire carrying an electric current is bent into a turn or loop, all the magnetic flux linked with the wire enters this loop at one face and leaves it at the other face. Consequently, that face of the loop from which the flux emerges must correspond to the north magnetic pole, and that at which it enters, to the south magnetic pole, of an ordinary bar magnet. This is illustrated in Fig. 76, both in the case of a permanent steel magnet, and of an active coil. In the case of a magnet, the flux is represented as coming out of the north pole, as indicated by the arrows, traversing the region or space outside of the magnet, re-entering the magnet at its south pole, and continuing through the body of the magnet to its north pole, thus completing the magnetic circuit. Similarly, in the case of an active loop, as shown, if the current circulates around this loop clockwise, as viewed by an observer at A, then

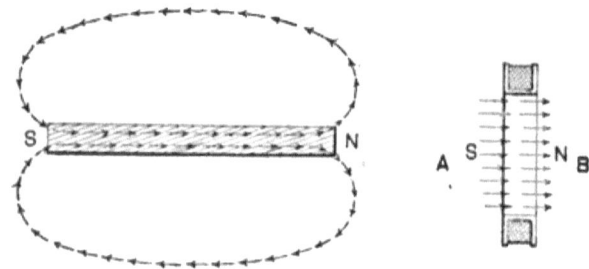

Fig. 76.—Diagram of Flux Produced by Permanent Magnet and by Coil of Active Conductor.

the flux will enter at A, and emerge at B, so that the face B, becomes a north pole, and A, a south pole, corresponding to the permanent magnet. In the case of an active loop, the flux paths form closed magnetic circuits as in the case of the magnet, although these are not shown in the figure. When the current in the conducting loop ceases, the magnetic flux linked with the loop entirely disappears.

Magnetic circuits are of three kinds; namely,

(1) Those in which all parts of the path of the flux are completed through air, or other non-magnetic material, such as wood, copper, glass, etc. Such a circuit is called a *non-ferric magnetic circuit*.

(2) Those in which all portions of the path of the flux are completed through iron or steel. Such a circuit is called a *ferric circuit*.

(3) Those in which parts of the circuital path of the flux are completed through iron or steel, and parts through air or other non-conducting material. This is called an *aero-ferric circuit*.

Non-magnetic circuits are formed by active conductors, such as wires, loops or coils carrying electric currents in the ab-

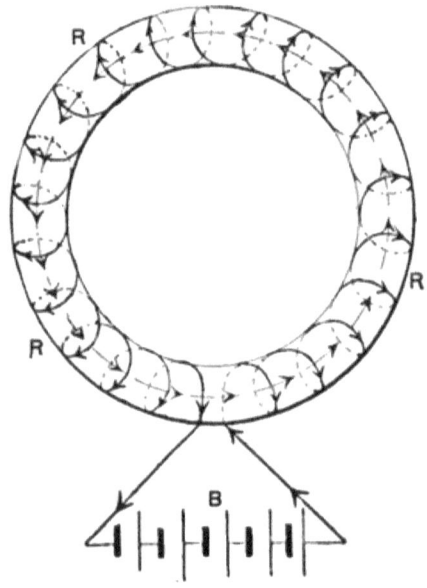

Fig. 77.—Ferric Magnetic Circuit.

sence of iron. An example of such a circuit is represented by the active coil shown in Fig. 77. Ferric magnetic circuits are less frequently met with, from the fact that the object of sending a magnetic flux through a circuit is to employ such flux in the operation of some mechanism, generally placed in a gap in the circuit itself. There is, however, a comparatively large class of apparatus called alternating-current transformers, which will be briefly explained later, and which almost always employ ferric circuits.

An iron ring, or core, wrapped with a coil of wire connected to the terminals of a battery, is an example of a **ferric** magnetic circuit. Such a magnetic circuit is shown in Fig. 77. Here all the flux due to the active conductors is entirely confined to the iron ring. A practical form of a ferric circuit is represented in Fig. 78, which represents an alternating-current transformer. The coil of active conductor is shown at AA, linked with a laminated or sheet iron core *BB*. Aero-ferric magnetic circuits are commonly observed in the case

Fig. 78.—Alternating-Current Transformer, Ferric Magnetic Circuit.

of permanent magnets. Thus the bar magnet shown in Fig. 76, has its magnetic circuit completed partly through the bar and partly through the air outside the bar.

A very common type of aero-ferric magnetic circuit is represented in Fig. 79, which shows an *electromagnet*, consisting essentially of a bar of soft iron AB, wound usually with a large number of

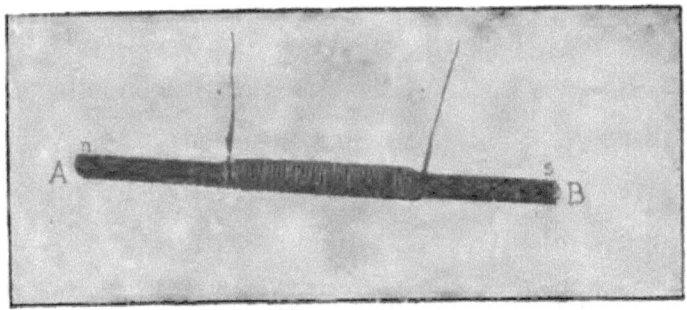

Fig. 79.—Bar Electromagnet.

turns of active conductor. Here the presence of the iron core causes the flux produced by the current passing through the coil, to be more powerful than that which the coil alone would produce. The polarity of the iron core AB, will, of course, depend on the direction of the current in the wire. If this direction be such that the flux enters the core at the end *B*, and leaves at the end *A*, then the north and south poles of the electromagnet so formed will be as marked in the figure. The introduction of the iron core, has not, therefore, altered the polarity produced by the helix, but it has greatly increased the quantity of magnetic flux, so that the magnet exerts a greater influence at a distance, and also a greater attractive power at its poles. Moreover, when the core is absent, the cessation of the magnetizing current is accompanied by a complete loss of the magnetic properties of the coil; that is to say, the copper wire forming the coil possesses no permanent magnetism. If, however, a core be present, the magnet does not immediately lose its magnetism on the cessation of the current. A certain amount of flux called *residual flux*, or *remanent flux* remains in the circuit. When the core of iron is very soft and carefully annealed, the amount of this *residual magnetism* is very small. When, however, the bar is formed of hard iron, a considerable portion remains on the cessation of the magnetizing current. In the case of a bar electromagnet, the magnetic circuit is largely formed of air, less than half of the circuit existing in the iron or steel.

If we bend the bar shown in the preceding figure, so as to bring the two poles nearer together, we get a form of electromagnet called the *horse-shoe electromagnet* in which the length of the air path is considerably reduced. Instead of actually bending the bar, it may, for purposes of convenience, be made in three separate parts as shown in Fig. 80, which is the form ordinarily given to an electromagnet. Here the magnet consists of two separate iron cores, connected together at the ends by a bar of soft iron called a *yoke*. Each of the two cores is provided with a magnetizing coil.

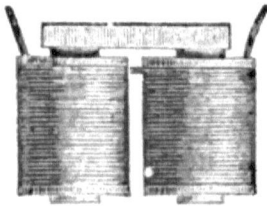

Fig. 80.—Electromagnet.

The value of the electromagnet depends largely on the fact that its core being made of very soft iron, possesses the property not only of greatly increasing the strength of the flux produced by the magnetizing coils, but also of readily losing nearly all its magnetism on the cessation of the magnetizing current; so that such a magnet when small can readily acquire and lose its magnetism, many times in a second.

Various forms are given to electromagnets according to the purposes for which they are designed. Where it is desired that an electromagnet should possess the power of attracting or repelling magnetic bodies at a considerable distance from its poles, the circuit is necessarily of the aero-ferric type, since the flux must pass for a considerable distance through air; but where it is desired that the magnet shall possess the power of holding heavy weights attached to its *armature*,—the name given to the bar of iron completing the magnetic circuit,—this circuit approaches more nearly to the ferric type.

A form of magnet capable of producing very powerful magnetic flux in the space between the poles is shown in Fig. 81. It consists of two powerful magnetizing

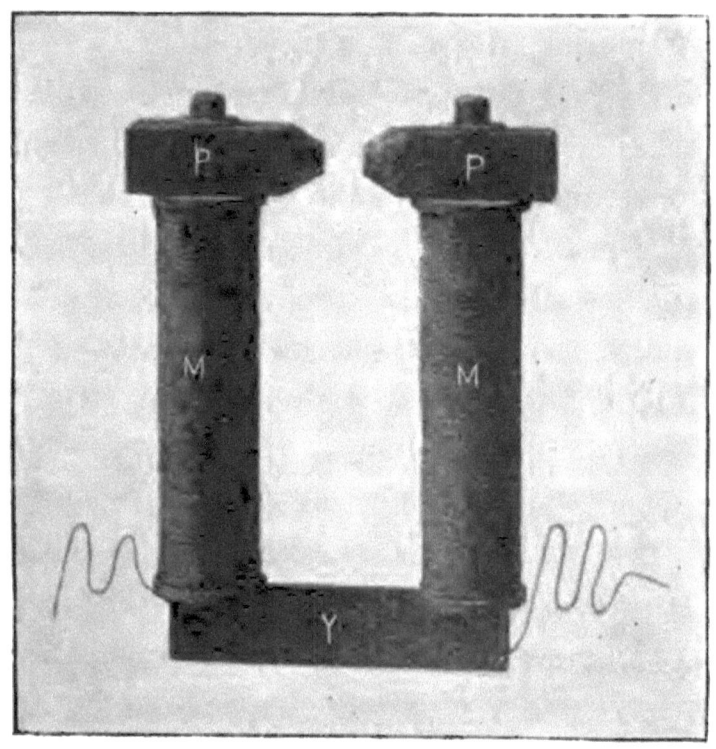

FIG. 81.—ELECTROMAGNET.

coils *M, M*, wound on iron cores, iron yoke *Y*, and iron pole pieces *P, P*. The power of an active coil to produce a magnetic flux is called its *magnetomotive force*, generally contracted M. M. F. This magnetomotive force is proportional to the number of turns of wire, and also to the current strength passing through the same. Consequently, if we add more turns to a coil, or increase the current strength passing through it, we will increase its M. M. F., and produce a greater magnetic flux through the circuit.

In every magnetic circuit the strength of the magnetic flux depends on two quantities; namely, the resistance opposing the magnetic flux, called the *magnetic resistance* or *reluctance*. A similar resistance to the passage of electrostatic and electric flux exists in the case of both the electrostatic and electric circuits. The value of the magnetic flux, like that of the electric flux may be expressed by the formula of Ohm's law, as follows; namely,

$$\text{Magnetic Flux} = \frac{\text{Magnetomotive Force}}{\text{Reluctance}},$$

so that if we know the magnetomotive force in a magnetic circuit, and the value of the magnetic resistance or reluctance, by dividing the former by the latter, we obtain the value of the magnetic flux.

As in the case of the electric circuit, special names are given to the unit values of these quantities. The units of *magnetomotive force* are called the *ampere-turn*, and the *gilbert*, the ampere-turn being greater than the gilbert in the ratio of approximately 5 to 4. By an ampere-turn is meant the amount of magnetomotive force, which is produced by a turn of wire carrying a current of one ampere; that is to say, if the magnetizing coils shown in Fig. 80

consisted of 200 turns in each spool, and if a current of five amperes passes successively through these spools, then the total M. M. F., urging the magnetic flux through the circuit, is 5 amperes × 400 turns = 2,000 ampere-turns = $\dfrac{2{,}000 \times 5}{4}$ = 2,500 gilberts, approximately. The amount of magnetic flux produced in the circuit by this M. M. F. will depend entirely upon the reluctance of the circuit. If the air-gap is large; *i. e.*, if the magnetic circuit contains a long air path, the magnetic resistance, or reluctance, of the circuit will be great, and the magnetic flux produced by the magnetomotive force will be comparatively small. If, on the other hand, the length of the air-path is small, the reluctance will be very small, and the amount of flux produced will be correspondingly great. This is for the reason that the reluctance of iron is very small as compared with that of air, provided that the iron is not saturated; *i. e.*, is not already conducting a large amount of flux per square inch or per square centimetre of cross-sectional area.

As in the electric circuit, the resistance of a wire depends upon its length and cross-sectional area, as well as on the nature of the material of which it is composed, so, in the magnetic circuit, the reluctance depends upon the length and area of cross-section of the circuit, and on the nature of the substance through which the flux is passing. In order to decrease the resistance of a wire, we may either decrease its length, or increase its area of cross-section. In the same way, in order to decrease the reluctance of a magnetic circuit, we may decrease its length or increase its area of cross-section. The *resistivity*, or resistance in a unit cube, varies markedly with the nature of the substance, **but does** not vary with the current strength passing through the material, provided the temperature is considered as remaining the same. In the magnetic circuit, the *reluctivity*, or reluctance in a unit cube (reluctance in one cubic centimetre measured between parallel faces) is practically the same for all substances other than the magnetic metals; in which the reluctivity is much lower.

Unlike the case of the electric circuit, the reluctivity varies markedly with the strength of the magnetic flux passing through the circuit. When the magnetic flux passing through iron is feeble, the reluctivity may be a thousand times less than that of air, while iron magnetically saturated; *i. e.*, carrying a very dense magnetic flux, has a reluctivity practically equal to that of air. The unit of reluctance is called the *oersted*, and is equal to the reluctance offered by a cubic centimetre of air, or more strictly of air-pump vacuum, measured between parallel faces, and is, therefore, nearly equal to the reluctance of a cube of glass, air, wood, copper, etc., measured between parallel faces. The reluctivity of air is, therefore, taken as unity.

The unit of magnetic flux is called the *weber*. One weber will flow in a magnetic circuit under a M. M. F. of one gilbert, through a reluctance of one oersted. If the reluctance of the magnetic circuit represented in Fig. 80, be 0.5 oersted, then since its M. M. F. has been assumed at 2,500 gilberts, the flux through the circuit will be $\frac{2,500}{0.5} = 5,000$ webers.

There are but two ways of varying the M. M. F. in a circuit; *i. e.*, by increasing the number of turns, or by increasing the current strength circulating in them; or, briefly, by increasing the number of ampere-turns.

It is necessary to draw a distinction between the total flux in a circuit measured in webers, and the intensity of the flux per unit of cross-sectional area; *i. e.*, per square inch, or per square centimetre; just as it is necessary to distinguish between the total current strength in an electric circuit, as measured in amperes, and the density of that current, as measured in amperes-per-square-inch, or per-square-centimetre, of cross-sectional area in the wire conveying the current. Since, in the case of the magnetic circuit, the introduction of iron is invariably attended by an increase in the magnetic flux, it is evident that by the introduction of a sufficiently great amount of iron, the amount of magnetic flux can be increased almost to any extent. Although this would necessitate a marked increase in the area of cross-section of the iron, yet the *flux density* per square inch, or square centimetre, may not be increased. Since soft iron practically saturates at an intensity of 19,000 webers-per-square-centimetre, and its reluctance near saturation rapidly increases, it is difficult to obtain at any portion of the magnetic circuit, intensities higher than 19,000 webers-per-square centimetre; *i. e.*, 19,000 gausses, the *gauss* being the *unit of magnetic intensity*, or the density of one weber-per-square-centimetre of perpendicular area of cross-section. The intensity of the earth's magnetic flux is, approximately, half a gauss, while the highest experimental intensity on record is 45,350 gausses.

It does not appear that magnetic flux produces any apparent physiological effects on the human body. The human body, containing in its composition no appreciable quantity of magnetizable material, has practically the same reluctivity as ordinary air; that is to say, the interposition of the human body in a magnetic circuit does not appreciably affect the distribution of the magnetic flux. For example, if a delicately suspended magnetic needle be deflected by a magnet placed at a certain distance from it, the direct interposition of the body of a person between the magnet and needle is not found to produce any appreciable effect, although if the same person wears, for example, an iron rimmed pair of spectacles, or carries a key in his pocket, the effect on the needle may be very marked. This is because the magnetic flux, acting on the needle, passes through the body of the person as readily as through the previously intervening air, but the magnetic influence of the iron rimmed spectacles, or the key, may have a powerful influence on a delicately suspended needle even though twenty feet away from it.

Not only is the reluctivity of the human body practically the same as that of other non-magnetic materials, but portions of the body subjected to powerful magnetic fluxes do not appear to have produced in them any appreciable physiological effects. Although experiments are still wanting concerning the physiological influence which long sustained powerful magnetic flux may exert, yet it has been shown that human beings and dogs subjected for many minutes to intensities of magnetic flux of about 2,500 gausses, and, therefore, about 5,000 times that of the earth's magnetic flux, have not experienced any influence that could be observed. Similarly, experiments made both with continuous and rapidly alternating magnetic fluxes have not shown any effect produced upon the circulation of the blood due to the iron it contains, upon ciliary or protoplasmic movements, upon sensory or motor nerves, or upon the brain.

It has been positively asserted that in a perfectly dark room certain individuals possess the power of observing faint luminous phenomena, around the poles of permanent or electro-magnets; that is, that these persons actually possess the power of being visually affected by magnetic flux. Investigations, however, have not only thrown doubt upon the original experiments, but repetitions of these experiments, with powerful electromagnets, have entirely failed to confirm the statements.

So far, therefore, as we know at the present time, it would appear that magnetic flux is absolutely without influence either upon the human body, or on any of its physiological processes, and that, consequently, if any therapeutic effects do attend the use of magnets, the causes must be of a psychic rather than of a physiological nature. It is to be remembered, however, that carefully conducted researches with very powerful magnetic fluxes may yet show lesser residual influences, which the experiments up to the present time have failed to bring to light.

But up to the present time experiments made on human beings have failed to establish any physiological effects whatever, even when such a delicate organ as the brain is placed in the direct passage of a powerful magnetic flux. When, for example, a person is placed with his head between the poles of a powerful dynamo-electric machine, from which the armature has been removed, so that the flux passes directly through the head, even prolonged exposure has failed to produce any observed effect either on the pulse or respiration, whether the magnetic flux was intermittent or was steadily maintained.

Or, take the case of a powerful electro-magnet, made by wrapping an iron cannon with a suitable magnetizing coil, and producing a flux sufficiently great to cause heavy iron bars or bolts to be sustained on the person of a soldier standing before the gun, as shown in Fig. 82. Under these

Fig. 82.—Magnetic Gun Attraction through a Soldier's Body.

circumstances, no sensations were experienced by the soldier other than those of pressure from the attracted masses of iron.

It would appear evident from the preceding observations that very little credence can be placed on the extravagant claims as to the curative power possessed by small magnets carried or worn on the body. The magnetic flux produced by such magnets is necessarily comparatively feeble, and if the more powerful fluxes before referred to failed to produce any appreciable physiological effects, there are no reasons for believing that these feeble fluxes can produce any marked effects unless that due to a feeble influence, long sustained. In the case, however, of most of the magnetic nostrums, for which curative effects are claimed, even the weak flux they produce usually fails to be properly directed, does not pass through any portion of the body, and can, therefore, have no physiological effect, except through the medium of the imagination.

CHAPTER IX.

INDUCTION OF E. M. F. BY MAGNETIC FLUX.

When a conducting loop is filled with, or emptied of, magnetic flux, electromotive forces are thereby set up or induced in the loop. This is called the *induction of E. M. F. by magnetic flux*. Four cases of such induction may arise; namely,

(1) Self induction.
(2) Mutual induction.
(3) Electro-magnetic induction.
(4) Magneto-electric induction.

We have seen that when an electric current circulates through a coil or loop, all the flux produced by the current is caused to enter the loop at one face, and to emerge at the opposite face. When a circuit is closed, so that the electric source begins to force electric currents through the circuit connected with it, some little time is required before the full current strength is established; so that, during this time, the magnetic flux that is passing through the loop is increasing in strength. Also, when the circuit is opened, some time is required for the current to entirely cease flowing through the circuit, and, during this time, the magnetic flux passing through the loop is decreasing. Therefore, both at the moment of making and breaking an electric circuit, a tendency will exist, if the circuit contains coils or conducting loops, for electromotive forces to be induced in the circuit. These E. M. Fs. continue only while the current strength is varying; as soon as the current strength in the circuit becomes constant, they disappear.

The amount of the E. M. F. induced at any moment of time in a conducting loop, by filling or emptying it with flux, depends upon the rate at which the loop is filled with, or emptied of flux. Suppose, for example, that 100,000 webers are passed through a loop, in, say two seconds of time: then if the rate at which this flux enters the loop is uniform, the E. M. F. generated in the loop will be maintained during the entire two seconds, and will be equal to the rate of entry in webers-per-second, or $\frac{200,000}{2} = 100,000$ webers-per-second = 100,000 units of E. M. F. The unit of E. M. F., the volt, has been so chosen that 100,000,000 webers, passing through the loop per second, generate one volt; so that this E. M. F. is $100,000 \div 100,000,000 = \frac{1}{1,000}$ volt. If, however, the 200,000 webers, above mentioned, entered the loop in say 0.01 of a second, the E. M. F. induced in the loop would be 200 times greater, or $\frac{200,000}{0.01} =$ 20,000,000 units of E. M. F. = 0.2 volts, but this E. M. F. would only last for the 1/100th of a second. When, therefore, a loop is filled with and emptied of a given number of webers of flux, the E. M. F. which will be produced in the loop depends entirely upon the time in which the filling and emptying takes place. If the filling takes place very suddenly, the E. M. F. will be powerful, but of very short duration. On the other hand, if the filling or emptying takes place slowly, the

E. M. F. will be correspondingly weaker, but longer sustained.

The direction of the E. M. F. induced by filling a conducting loop with flux, is opposite to that induced by emptying the same loop of flux. The direction of the E. M. F. induced by filling a conducting loop with flux, is readily remembered by the following rule:

Regarding the loop as the face of a watch, held in front of the observer, then if the flux passes through the loop in the *same* direction as the light passing from the face of the watch to the observer's eye, the E. M. F. induced in the loop will have the *same* direction as that of the hands of the watch.

Some general idea concerning the manner in which E. M. F. is generated in

a loop by the passage of magnetic flux
through it, may, perhaps, be obtained from

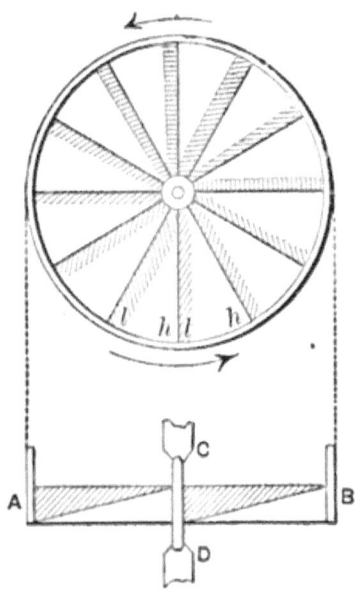

Fig. 83.—Mechanical Model Having Analogies with
Electric Circuit.

the mechanical model shown in Fig. 83.
A cylinder AB, pivoted upon a vertical
axis CD, mechanically represents a conducting loop of wire. The cylinder is con-

nected with the axis by a number of radial spokes in the form of fan-blades, so that, if a stream of liquid, such as water, be poured through the cylinder or loop from above, the impact of the water on the blades will cause the cylinder to rotate in a direction opposite to that of the hands of a watch. If, however, the water be forced upward through the loop, its impact will cause the cylinder to rotate in the opposite direction. If a given number of gallons of water be passed through the cylinder, the driving impulse communicated to it will depend upon the time during which the water passes. If the water be delivered in a brief time, its rate of passage through the loop will be great, and the driving impulse communicated to the cylinder will be great, though of brief duration. If, on the other hand, the time during which the water passes through the loop be considerable, the driving impulse exerted on the cylinder will be prolonged, but correspondingly feeble. It will be observed that the driving impulse or force in this mechanical analogue stands for electromotive force in the electrical case. When the water is first poured through the cylinder, the inertia of the cylinder will prevent it from being immediately set in motion. Similarly, when the water has ceased to pass, the motion of the cylinder, owing to inertia, does not immediately cease. In the electric circuit, this corresponds to the effect of self-induction; for, the effect of pouring flux into a loop is to induce in the loop an E. M. F., and the effect of the current so set up, is to produce in the loop a flux opposite to that of the inducing flux, producing thereby a C. E. M. F. retarding the development of the electric current. On the other hand, when the flux has filled the loop, the current does not immediately cease flowing, being prolonged by the action of the flux set up by the current; in other words, the loop acts as though it possessed electrical inertia.

When a number of turns are connected in series, as, for example, in the case of the coil of conducting wire shown in Fig. 77, the effects produced by each turn are added, so that the coil has induced in it an E. M. F. proportional to the number of its turns. When a conducting coil, containing many turns, has its terminals connected to a voltaic battery, some little time elapses before the full current strength is established in the circuit. The reason is to be found in the *C. E. M. F. of self induction* of the coil. Similarly, when the circuit of this coil is opened, the current does not instantly cease flowing through it, since the emptying of the coil of the flux, produces in it an E. M. F. which establishes in the coil a current in the same direction as that sent through it by the battery. In other words, the E. M. F. produced in the coil by self-induction, at the moment of making, tends to *oppose* the establishment of the current, and that produced at the moment of breaking, tends to aid the passage of the current.

When the circuit of an electric source, such as a voltaic battery, is opened, a minute spark is frequently visible at the point of opening. If, however, a coil of many turns of wire be contained in the circuit, the spark upon opening the circuit will, probably, be much greater, and a distinct shock may be felt under favorable conditions by the person opening the circuit. This spark is due to the self-induction of the circuit. The current which has passed through the circuit has produced a magnetic flux linked with the turns; *i. e.*, the loops in the coil or coils of wire. On the opening of the circuit, this current is suddenly interrupted, and the flux, rapidly disappearing from the coils; *i. e.* pouring out of them, induces a brief but powerful E. M. F., which, acting in the same direction as the current, tends to prolong it.

If two coils A and B, Fig. 84, connected in separate circuits, are placed side by side, and an electric current be sent through one, say A, the passage of this current will produce a magnetic flux, part of which will pass through B. During the process of filling B, with this portion of A's flux, an E. M. F. will be set up in each turn of B, equal, at any moment, to the rate at which

the flux is entering in webers-per-second;
or, expressed in volts, to the rate at which
the flux is entering in hundred millions of
webers-per-second. As soon as the cur-

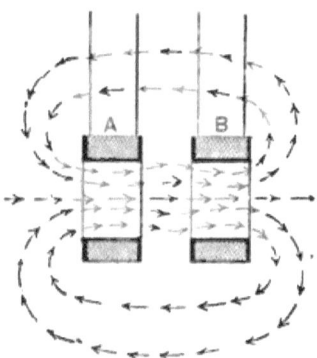

Fig. 84.—Diagram Illustrating Mutual Induction.

rent in A, becomes stationary, the flux
through B, due to this current, is also
stationary, and, consequently, no further
E. M. F. is induced in B. If, however, the
current strength in A, diminishes, its flux
through B, will be correspondingly diminished, and an E. M. F. will be induced in

B, in the opposite direction to that originally produced, and equal in volts to the rate of emptying in millions of webers-per-second. When the flux through B, due to the current in A, has entirely disappeared, the E. M. F. induced in B, has also disappeared. If there be 100 turns of wire in the coil B, the E. M. F. induced in the coil will be 100 times as great as if it consisted of a single turn, assuming that the same quantity of A's flux passes through all of B's turns alike. This inductive influence extending from one coil to another, whereby a current in one circuit induces an E. M. F. in another circuit, is called *mutual induction*. An example of mutual induction can be shown by means of the apparatus represented in Fig. 85, in which A, represents the *inducing coil* ; *i. e.*, the coil in which the current flows; and B, the coil in which the current is induced. Or, as they are generally called, *A*, is the *primary coil* and *B*, is the *secondary coil*. If the terminals of the primary coil *A*, be connected with

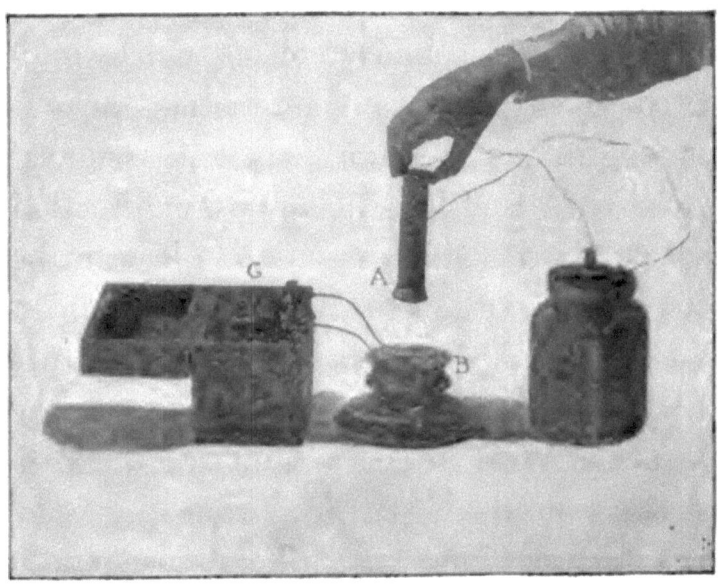

Fig. 85.—Mutual Induction.

the voltaic cell *C*, as shown in the figure, and the terminals of the secondary coil *B*, be connected with an ammeter, or galva-·nometer, *G*; then, as soon as the current is established in A, no current will be induced in B, as long as A, remains at rest. If, however, A, be moved either toward or from B, currents will be produced in the secondary coil, as will be indicated by the galvanometer, the current passing in one direction, when A, is moved toward B, and in the opposite direction when A, is moved from B. It can be shown that the current induced in a secondary coil is induced in the opposite direction to that in its primary, on the approach of A to B, and in the same direction, on its withdrawal from B. The two circuits A and B, although electrically disconnected, are connected magnetically by the flux permeating the space between them, and the E. M. F. of mutual induction is caused by the flux proceeding from the primary coil being carried toward or from the secondary coil, during its motion, so as to cause the secondary coil to be filled with more or less flux.

That mutual induction may take place between stationary primary and secondary coils, may be experimentally demonstrated. For example, if as in Fig. 86, the primary coil A, is fixed at a constant distance from B, then on completing the circuit of the primary coil, by closing the switch S, while the current is increasing in the primary, the magnetic flux produced by it passes through the conducting loops on the secondary coil C, thereby inducing an E. M. F. and establishing a current, as is shown by the galvanometer G.

The distinction between electro-magnetic and magneto-electric induction is seen in Fig. 87, where the motion of the magnet M, into or out of the coil of wire, produces electromotive forces in the coil C, as shown by the galvanometer G. When the magnet is thrust into the coil, the

Fig. 86.—Mutual Induction.

galvanometer indicates a temporary current in one direction, and, on its withdrawal from the coil, it shows a current in the opposite direction. The introduction of the south pole into the coil produces the same direction of current as the withdrawal of the north pole. Here, as in the other instances, E. M. Fs. are

Fig. 87.—Magneto-Electric Induction.

induced by the passage of magnetic flux through the coil; the flux produced by the magnet, being advanced or moved so as to pass through, or link with, the turns in the secondary coil.

A form of apparatus for producing E. M. Fs. by magneto-electric induction is

ELECTRO-THERAPEUTICS. 239

represented in Fig. 88. A permanent horse-shoe magnet, *MM*, is supported in

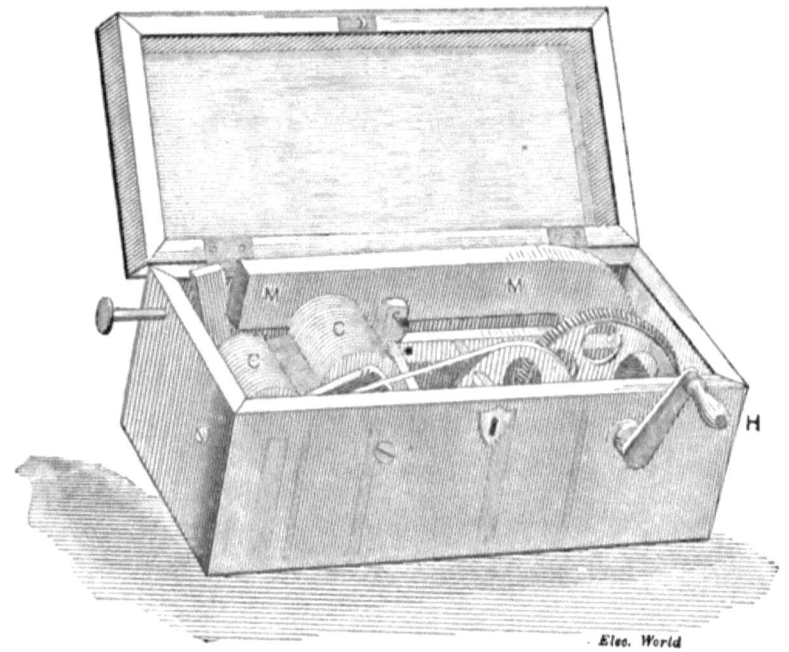

FIG. 88.—MAGNETO-ELECTRIC GENERATOR.

a vertical position, and two coils of fine insulated wire *CC*, are supported on a horizontal axis, in such a manner as to be capable of rotation by the turning of the handle *H*, the rotary speed of the coils being made greater than the rotary speed of the handle, by the interposition of suitable multiplying gear. The coils are wound on cores of soft iron, which are connected by a soft iron yoke *y*. The ends of the cores revolve in close proximity to the poles of a permanent magnet, leaving a small air gap or clearance of comparatively small reluctance. When the two coils stand vertically, the flux from the magnet passes through the air gap, the cores and their connecting yoke, thereby filling all the turns of wire wound upon the core. An E. M. F. will be induced in each turn equal in volts to the rate of filling it with flux, in hundred millions of webers-per-second ; and, since all the turns in each coil, and the two coils themselves, are connected in series, the total E. M. F. will be correspondingly multiplied.

The condition of affairs in the preceding machine, is represented in Fig. 89, where

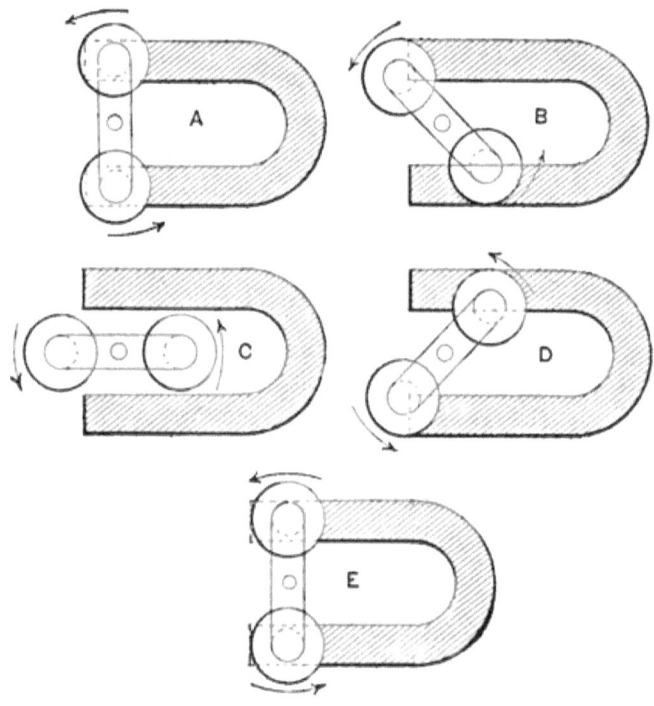

Fig. 89.—Diagram Representing Changes in the Magnetic Circuit of Magneto-Electric Generator.

the coils are shown at *A*, as being immediately opposite to the magnet poles, and in such a position as to be filled with flux, so that they cannot receive any further increase of flux by a further rotation in either direction.

At B, the coils are leaving the pole pieces, so that the reluctance in the magnetic circuit is increasing, and the magnetic flux, which passes through the cores of the coils, is diminishing. In other words, the coils are becoming emptied of the flux they contain. An E. M. F. is, therefore, induced in them.

At C, the coils are completely emptied of magnetic flux, and, therefore, have no E. M. F. At D, the coils are being filled with flux, but in the opposite direction to that which exists at A. Consequently, the E. M. F. induced has the opposite direction to that induced at B.

At E, the coils are full of flux in the opposite direction to that at A, and the E. M. F. in the coils will have ceased.

It will, therefore, be evident that during any half revolution, as from A to E, the E. M. F. induced in the coils has made a single alternation or reversal; and that during the next succeeding half revolution, in which the coil returns to the position A, the E. M. F. induced will be of the same magnitude as above pointed out, but in the opposite direction.

The revolving coils, therefore, generate alternating currents in the circuit connected with them, the E. M. F. being alternately in opposite directions, during successive half revolutions. One complete revolution of the coils produces one complete *double alternation*, or *cycle* of the E. M. F. and electric current, consequently, the *frequency* of the alternating currents produced; *i. e.*, the number of complete double alternations, or cycles per second, is equal to the number of

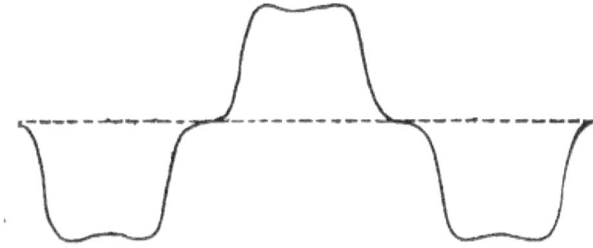

FIG. 90.—DIAGRAM OF A POSSIBLE WAVE FORM OF MAGNETO-ELECTRIC GENERATOR E. M. F.

revolutions made by the coils per second. The alternating E. M. F. produced by this machine might be represented diagrammatically in Fig. 90. The exact wave form, in each case, would depend upon the shape of the poles and of the iron cores. If, however, a commutator be employed on the armature, as shown in Fig. 91, whereby at each half revolution the connections of the coils with the external circuit is reversed, the current produced by the alternating E. M. F. will be unidirectional in the external circuit.

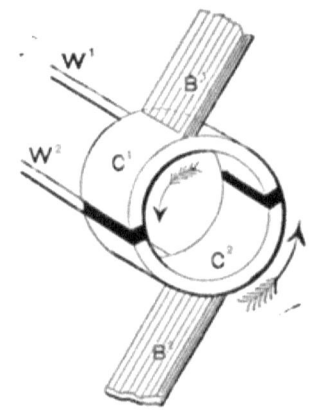

Fig. 91.—Diagram of Two-Part Commutator.

Fig. 92, represents the corresponding form of pulsating E. M. F. wave produced in the external circuit when the commutator is employed. It will be seen that the E. M. F. is now always above the line. If the E. M. F. were reversed, the waves might be represented as being entirely below the line.

Fig. 93, represents a form of magneto-electric machine, the current from which is capable of lighting a small incan-

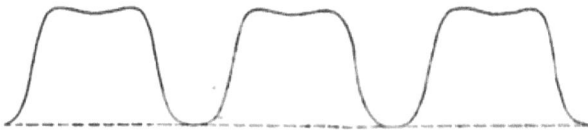

Fig. 92.—Diagram of a Possible Wave Form of Magneto-Electric Generator E. M. F. (when a Commutator is Employed).

descent lamp. If an *alternating magneto-electric generator*, that is a magneto-electric generator not employing a commutator, is connected to the body of a patient by suitable electrodes, alternating electric currents will pass through the body. If, however, a commutator be employed, and the currents be of the wave form shown in Fig. 92, the physiological effects will be somewhat different. The type of current of Fig. 93, possesses polar properties; *i. e.*, possesses the characteristics of unidirec-

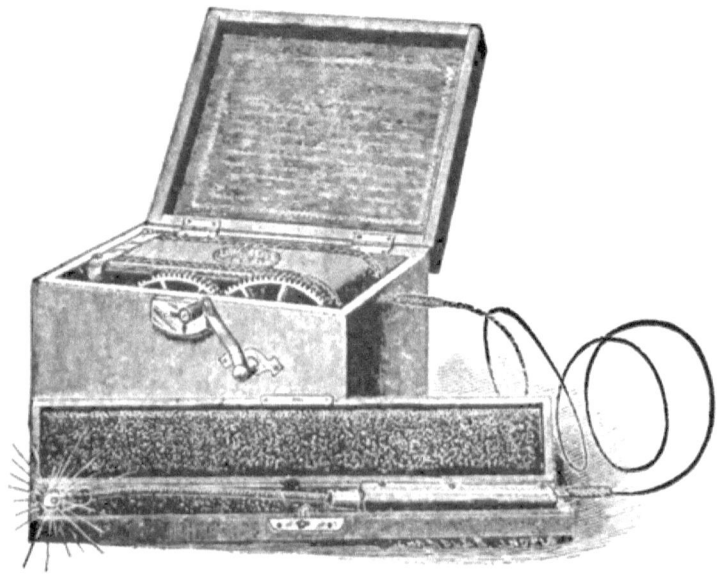

FIG. 93.—MAGNETO-ELECTRIC GENERATOR.

tional currents, while symmetrical alternating currents do not possess these properties, since the polar effects produced by one wave, are neutralized by the following wave in the opposite direction.

CHAPTER X.

THE MEDICAL INDUCTION COIL.

The *medical induction coil*, generally called the *faradic coil*, is very frequently employed in electro-therapeutics. It consists essentially of means whereby E. M. Fs. are induced by mutual induction, and, consequently, of a primary and a secondary circuit. A simple form of induction coil is represented in Fig. 94, where the terminals of the primary circuit are shown at P, P, and those of the secondary circuit at S, S. Fig. 95, shows a similar coil in longitudinal section. An inspection of the latter figure will show that the primary coil P, consists of a comparatively short length of fairly coarse wire, wrapped around a hollow bobbin. The secondary circuit generally consists of a greater length of finer fire wrapped either directly over the secondary, or on a hollow bobbin capable of being moved over the

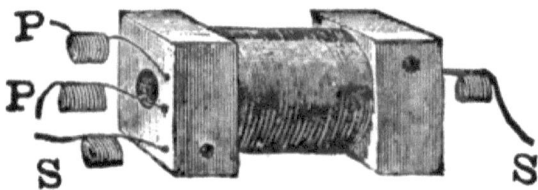

FIG. 94.—SIMPLE FORM OF INDUCTION COIL.

primary. When the current strength in the primary circuit is varied; *i. e.*, when either an alternating or a pulsatory current is sent through the primary, an alternating E. M. F. is induced in the secondary circuit by the influence of mutual induction.

The amount of E. M. F. induced in the secondary circuit depends upon the number of turns in its coil and the rate at which the magnetic flux fills and empties these turns. The induced E. M. F. does not depend upon the number of yards or feet of wire in the secondary coil, except in so far as a greater length of wire pro-

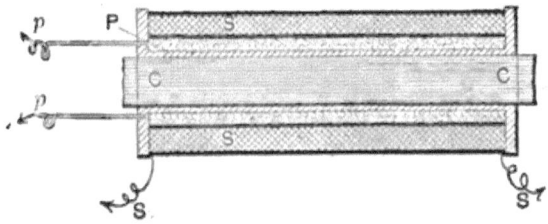

Fig. 95.—Section of Simple Form of Induction Coil.

vides a greater number of turns in the coil. If we double the number of turns in the coil without altering in any way the amount of flux which passes through each turn, we double the number of volts induced therein; whereas, if we double the number of feet or yards in the secondary coil, we do not necessarily, and in point of fact very rarely, double the number of turns, and, therefore, the number of volts, because the average length of turn in each successive layer increases. If we double the rate at which the flux threads or links with the turns of the secondary coil, we double the E. M. F. induced in the coil.

An increased rate of filling and emptying conducting loops or turns with flux can be obtained in one or both of two ways; viz.,

(1) By causing the same flux to fill and empty the loop a greater number of times per second; *i. e.*, increasing the frequency of oscillation of the flux in the magnetic circuit; and,

(2) By increasing the amount of magnetic flux in the circuit without increasing the frequency of oscillation, so that more flux enters or fills the coils at each alternation.

Consequently, for a given primary and secondary circuit, with a given geometrical relationship between them, we can only increase the E. M. F. in the secondary circuit either by increasing the magnetic flux, or by increasing the frequency of flux oscillation, or both.

In order to increase the frequency of flux oscillation, we require to increase the frequency of the primary current. On the other hand, in order to increase the total amount of flux we must either increase the M. M. F. of the primary circuit, or diminish the reluctance of the magnetic circuit; that is to say, we must either employ more ampere-turns at the same frequency, or employ such a form of iron core in the primary coil as will increase the magnetic flux from a given M. M. F. by diminishing the magnetic resistance of its circuit.

The frequency required for the primary circuit may be obtained either by an alternating, or by a continuous, but pulsating current. The ordinary faradic coil

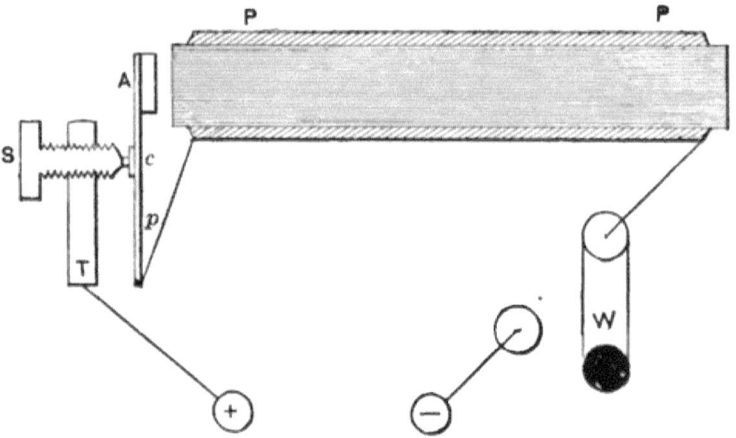

Fig. 96.—Primary Connections of Medical Induction Coil.

only employs the latter, the connections being represented in Fig. 96. As soon as the circuit is closed at the switch W, the current flows through the primary coil of the instrument, into the spring p, through the contact c, and the screw stud S, in the support T. The M. M. F. of this current produces a magnetic flux passing through the core, and through the air outside, in circuital paths. This flux being produced within the magnetic circuit, sets up a C. E. M. F. of self-induction, tending to retard the development of both flux and current in the primary coil, so that the primary current does not instantly reach its full strength, but rises comparatively slowly to a maximum. As soon as sufficient magnetic flux has been produced in the magnetic circuit, to move, by magnetic attraction, a soft iron armature A, supported at the extremities of the spring p, the spring is forced to leave the contact c, and thus open the circuit. As soon as the circuit opens, the current strength would immediately fall to zero, but for the fact that the magnetic flux in the circuit, being unsupported by M. M. F., rapidly disappears, so that the primary loops are rapidly emptied of flux and become the seat of an E. M. F. tending to prolong the current. This E. M. F. is comparatively powerful, owing to the rapid rate at which the flux is emptied; in fact, it may be sufficiently great to cause a spark to form between the spring and the contact c. In this manner both the making and the breaking of the primary circuit produce E. M. Fs. in the circuit; that on making, tending to oppose the establishment of the current, and that on breaking, tending to oppose its cessation.

On the closing of the primary circuit, the time occupied in producing the full flux is comparatively great; that is to say, it may amount to, perhaps, the one hundredth of a second; the loops do not, therefore, fill so rapidly; but when the circuit is opened at the contact c, the flux is necessarily withdrawn with great rapidity, and the E. M. F. induced at breaking, owing to this greater rate, is much in excess of the E. M. F. induced at making. The value of the E. M. F. on breaking will, therefore, be increased by any cause which will tend to diminish the time required for the complete cessation of the primary current. The spark which bridges the space between the contact point and the spring has, therefore, the effect of prolonging the current, since it provides a path, of heated air, through which the current may flow, even when the contact is broken. Consequently, any device which will stop the spark will result in the production of a higher E. M. F. of self-induction in the primary coil. This is sometimes effected by introducing a condenser into the primary circuit, with its terminals connected in shunt to the contact. As soon as the contact is broken, the current instead of following through the air in a spark, rushes into the condenser and charges it. As soon as the condenser is charged, the current ceases very suddenly, and is then unable to jump across the interval of air which has become interposed between the contact point and the spring. The result is, therefore, that the current in the primary coil, being very suddenly arrested in the condenser, the rate at which the loops are emptied of flux is greatly increased, and the E. M. F. induced in the primary circuit is correspondingly increased.

So far we have only considered the primary coil as though the secondary coil were entirely removed from it. We may now consider the effects that are produced in the superposed secondary coil. If the terminals of the secondary coil be opened so that its external resistance is practically infinite, it can, therefore, send no current. The E. M. F. induced in the secondary coil is the exact counterpart of that which is induced in the primary coil, except, that having more turns, the secondary E. M. F. is correspondingly greater. This is on the assumption that all of the magnetic flux threading through the primary coil, also threads through the secondary. On closing the primary circuit, the entrance of magnetic flux through the primary and secondary loops together causes E. M. Fs. to be induced in both coils. This E. M. F. is a C. E. M. F. in the primary circuit, since it acts against the E. M. F. impressed upon it, but it is the only E. M. F. which appears in the secondary circuit. If, for example, the primary coil consists of 100 turns of wire, and has induced in it a C. E. M. F. commencing at, say one volt, as shown in Fig. 97, this C. E. M. F. dying away, along the curved line bcd; then, if the secondary

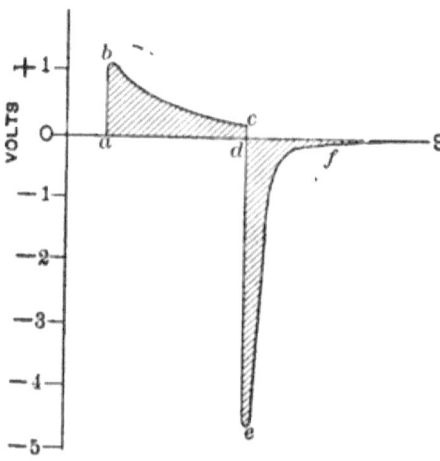

Fig. 97.—Diagram of Primary Induced E. M. Fs.

coil consists of 5,000 turns, the E. M. F. induced in this secondary coil will be represented by the same curve on a scale 50 times as great; that is to say, commencing at 50 volts, instead of at 1 volt.

The moment that the primary circuit is broken at the contact c, the E. M. F. induced in the primary circuit, instead of being 1 volt reducing to almost zero in the hundredth of a second, may be 5 volts, reducing to zero in, perhaps, $\frac{1}{1,000}$th of a second, as is represented in the Fig. 97, by the curve def. Similarly in the secondary circuit, the E. M. F. induced will be represented by the same curve on a scale 50 times as great, and will commence at 500 volts, reducing rapidly to zero. The E. M. F. of such a faradic coil is, therefore, obviously alternating, but dissymmetrical in character, the wave at breaking being much greater in amplitude, but correspondingly more brief than the wave at making.

If the secondary coil be moved away from the primary coil, as in instruments of the Dubois-Raymond type, one of which is represented in Fig. 98, where two secondary coils are shown with the primary partly inserted in one of them,

FIG. 98.—INDUCTION COIL, DUBOIS-RAYMOND TYPE.

the E. M. F. of self-induction, in the primary circuit, will not be influenced, but the E. M. F. of mutual induction in the secondary coil, will be reduced, because the amount of flux linked with the secondary turns is reduced, and finally, when the coils are entirely separated, the amount of flux from the primary coil, which remains linked with the secondary coil, is so small, that its filling and emptying produces an inappreciable induced secondary E. M. F. Consequently, in the use of the Dubois-Raymond type of coil, when it is desired to increase the secondary E. M. F., the primary coil is pushed further into secondary coil, or *vice versa*, so that the flux produced in the magnetic circuit may link with more and more turns in the secondary coil. In no case, can the E. M. F. in a secondary coil be other than increased by bringing the primary and secondary coils closer together, so long at least as the secondary circuit is open.

We have hitherto considered the secondary coil as being open, and have seen that in such cases, although E. M. Fs. are produced in the secondary, its presence has no appreciable effect upon the nature of the phenomena that occur in the primary circuit. As soon, however, as the terminals of the secondary coil are connected to an external circuit, so that a secondary current can be produced in such circuit, then effects are produced, which are superposed upon those already existing; namely, secondary currents flow. These secondary currents produce a M. M. F. in the secondary coil, and a magnetic flux through the secondary coil, part of which passes through the primary coil. This secondary flux, filling and emptying the secondary loops, sets up a C. E. M. F of self-induction in the secondary circuit tending to oppose both the development and the extinguishment of each current wave. Moreover, the flux from the secondary circuit, passing through the primary coil, induces in it an E. M. F. by mutual induction, and disturbs the flow of current strength through the primary coil. It is to this disturbance, and to the flow of primary current against the mutually induced C. E. M. F., that the energy is obtained from the primary circuit which appears in the secondary circuit. The amount of reaction taking place from the secondary back into the primary circuit, depends upon the strength of the secondary currents, which in its turn depends upon the resistance of the secondary circuit as composed partly of the resistance in the secondary coil, and partly in the external resistance, and also upon the self-induction, or, as it is usually called, the *inductance* of the secondary coil already mentioned. The greater the current strength in the secondary circuit, the greater the mutually induced C. E. M. F. in the primary circuit, and the greater the reactive disturbance in magnetic flux and in any current strength there existing.

Although, when the secondary circuit is open, its E. M. F. at the peak of the wave on breaking, may measure 200 or 300 volts, yet on closing the secondary circuit, through a comparatively low resistance, the E. M. F. at the secondary terminals may only be 2 or 3 volts. It is evident, that if the secondary terminals be short circuited by a stout piece of wire having negligible resistance, the E. M. F. at secondary terminals would be 0 volts, but even when the secondary external resistance is say, 1,000 ohms, the E. M. F. available at secondary terminals is always very small in the medical induction coil. This is both for the reason that, owing to the reaction of the secondary M. M. F. of the primary circuit, the E. M. F. in the secondary coil is reduced, and because such E. M. F. as remains, so soon as it tends to produce such a current in the external circuit, as would flow in accordance with Ohm's law, is met by a powerful C. E. M. F. due to its self-induction. This C. E. M. F. is most powerful when the wave of induced E. M. F. is most abrupt, since, at such times, the inrush of magnetic flux into the secondary coil from its own rising current is greatest. The result is that the drop of pressure in the coil is always very great, owing to its large resistance and large inductance; *i. e.*, its large number of turns and capability of producing C. E. M. F. by self-induction. The current which can flow in the secondary circuit is, therefore, not only comparatively feeble, but is also much less abrupt in wave character than the induced E. M. F. would lead one to expect.

The core employed in the medical induction coil almost invariably consists of a bundle of soft iron wires. A solid rod of soft iron cannot be efficiently employed, because, under the influence of the rapidly oscillating magnetic flux through such a rod, electric currents would be produced around it in such a manner as to oppose the development of the magnetization; that is to say, the external shells of the rod would act as closed secondary coils, tending to check and react upon the primary current and wastefully expend energy. By employing a bundle of fine, soft iron wires, all these induced eddy currents are reduced to negligibly small strengths

If the core, instead of being limited to the interior of the coil, be brought around, so that its extremities shall meet, and the core thus entirely surround the coil, the magnetic flux, produced in the magnetic circuit by a given primary M. M. F., will be much greater. It might, therefore, at first sight be supposed that such an arrangement would increase the induced secondary E. M. F.; but while the flux would certainly be increased, the residual magnetism in the ring of iron thus formed would be so considerable, that the rate at which flux could be caused to oscillate in the ring would be reduced in a greater proportion than the total flux could be increased, so that the employment of such a ring would be detrimental. If the primary current, instead of being pulsatory or unidirectional, were alternating, its influence in eliminating the residual magnettism, at each reversal, would be much greater, and the benefit of a completely ferric circuit would be proportionate.

The rate of vibration of the spring contact has an important bearing upon the action of the apparatus. The frequency of the primary pulsating current depends upon the natural frequency of vibration of the spring, which in turn depends upon its length, breadth, thickness, elasticity and the weight of the armature it supports at its free end. The lowest tone, which the spring vibrator will emit, corresponds to the slowest natural rate of vibration at which the spring will vibrate as a whole. By causing the contact to approach the core, so as to reduce the range of vibration, the spring is aided in producing *overtones;* i. e., vibrations set up from the contact point as a node, and in some cases the vibration of the spring is composite, being partly performed as a whole, and partly in segments, having the contact point as node.

The effect of increasing the frequency of vibration in the first instance is to increase the E. M. F. induced in the primary and secondary coils by self and mutual induction, owing to the greater rapidity with which the flux is compelled to enter and leave the coils. On the other hand, when the rapidity of vibration exceeds a certain value, depending upon all the electric conditions of the circuit, the time in each pulsation, during which contact is maintained at the vibrating spring, is so short that the entering primary current is greatly reduced in strength, since it has not time to overcome the C. E. M. F. of self induction. On this account the primary current waves are reduced in amplitude, and the M. M. F. is correspondingly reduced, thereby restricting the development of flux. When, therefore, this particular speed of vibration has been obtained, the gain in secondary E. M. F., by reason of the higher frequency, is offset and neutralized by the loss of total flux produced in the magnetic circuit. Moreover, the greater the rapidity with which the secondary waves of E. M. F. are induced, the more rapid will be the waves of M. M. F. and flux in the secondary coil, and the greater the secondary C. E. M. F. of self-induction tending to oppose the development of abrupt changes in the secondary current strengths. Although the effect of these separate influences depends upon the particular proportions of each induction coil, yet, broadly speaking, the effect of increasing the frequency of vibration in the contact spring may be expressed as follows. In all cases the frequency in the secondary circuit increases in conformity with the increase in the primary. Up to a certain frequency both the E. M. F. and the current strength in the secondary circuit are increased; beyond this frequency, the E. M. F. in the secondary circuit is not increased, but the wave type of secondary current becomes modified. The more rapid the vibration, the smoother and less abrupt the current waves produced, particularly, when long fine-wire secondary coils are employed, owing to the powerful C. E. M. F. set up by rapid fluctuations, and the throttling effect thus produced upon all abrupt variations of current strength.

In order, therefore, to produce in the secondary circuit abrupt waves of the type diagrammatically shown in Fig. 97, the secondary circuit should have a comparatively small number of turns of wire, placed as close as possible to the primary coil, so that the throttling effect of self-induction in its own turns may be avoided,

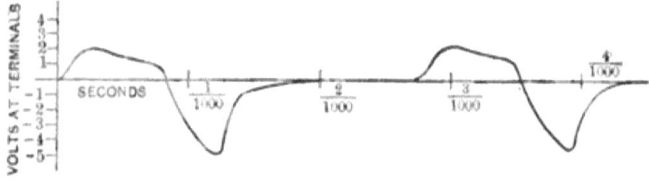

FIG. 99.—TYPE OF SECONDARY INDUCED E. M. F. AT HIGH FREQUENCIES UNDER LOAD.

and the frequency of vibration comparatively small. On the other hand, in order to produce in the secondary circuit, smooth, less abrupt current waves such as those shown in Fig. 99, long, fine-wire coils, with considerable inductance, should be employed. Moreover, it is not necessary that all the turns should be linked with the primary coil; that is to say, the secondary E. M. F., produced by mutual induction, can be induced in part of the secondary coil. The remainder of the secondary coil, being unacted on, will serve to choke or throttle the sudden variations in the secondary current by its inductance. This effect will be enhanced by making the frequency conveniently great. It is to be remembered, however, that in no case can either the E. M. F. wave, or the current wave, in the secondary circuit be made symmetrical, the wave on breaking being always steeper than that on making.

The usual method of altering the frequency in the primary circuit, consists in advancing the spring contact, so as to increase the tension of the vibrating spring, and reduce its range of vibration. The range of frequency obtained in this way is

ELECTRO-THERAPEUTICS. 275

comparatively small, as it is rarely that the frequency can be doubled in this manner. A device sometimes employed in connection

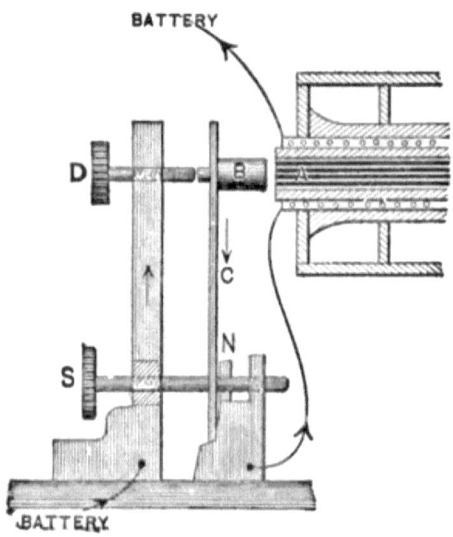

FIG 100.—ADJUSTABLE VIBRATOR FOR FARADIC COILS.

with larger induction coils is shown in Fig. 100, where the screw S, and the nut N, serve to tighten the vibrating spring and thus vary its tension. The frequency obtained from an ordinary spring vibrator is from 150 to 300∼; *i. e.*, 150 to 300 complete periods per second. For higher frequencies a *ribbon vibrator* is sometimes employed, such as shown in Fig. 101.

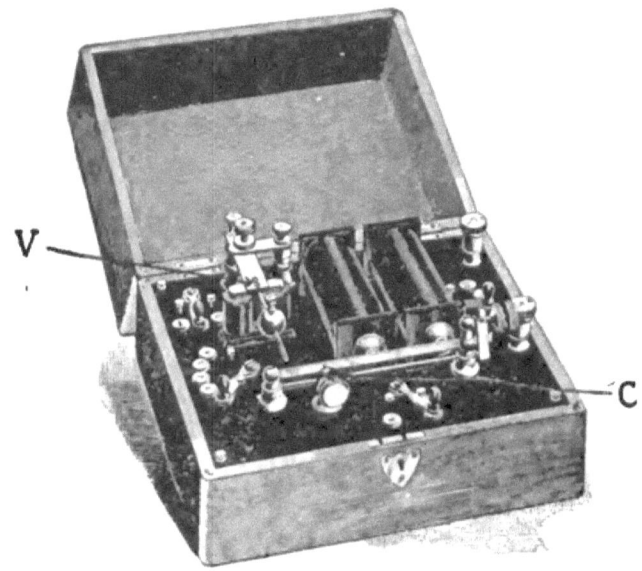

Fig. 101.—Induction Coil with Ribbon Vibrator.

In a ribbon vibrator the contact screw C, presses upon a horizontal steel ribbon, at a point which is about one quarter of the length of the ribbon from the fixed support on the left. The ribbon can be tightened by the thumbscrew on the right hand pillar or support. By varying the tension on the ribbon, a considerable range of frequency of vibration can be obtained. The induced secondary currents obtained from such an apparatus, with a long, thin-wire coil, have considerable frequency, small strength, and comparatively smooth, wave character. For abrupt, powerful, secondary currents, a slow speed vibrator is represented at V, in the same figure. It consists of a vertical electromagnet, which attracts an armature of soft iron, attached to a horizontal steel spring, and is loaded with a spherical weight, capable of being clamped at varying distances along the rod attached to the moving system. By clamping the weight near the free extremity of the rod, the slowest vibrations are obtained; while by clamping it near to the armature, the speed of vibration is increased. The secondary coil employed with the slow vibrator is usually of fewer turns and

Fig. 102.—Induction Coil with Rapid Interrupter.

coarser wire, offering a smaller resistance and a considerably smaller inductance.

Fig. 102, represents a type of medical induction coil in which the frequency is varied by means of a small electromagnetic motor MM. This motor drives three wheels 1, 2 and 3, upon the peripheries of which are a series of contact surfaces, which are pressed upon by spring contacts. By the insertion of one of these wheels in the primary circuit of the induction coil, the frequency of pulsation of the primary current can be varied within wide limits, and a high frequency attained. The secondary coil C, can be moved toward or from the primary coil within it, by turning the screw S.

Another method of varying the induced E. M. F. in a secondary coil, without varying the frequency, is by the use of a metallic tube inserted between the core and the primary coil, between the primary and secondary coils, or over the secondary coil. The second method is represented in Fig. 103, where the handle H, is attached to one extremity of a brass tube, inserted

between the primary and secondary coils
of the medical induction coil, which in this

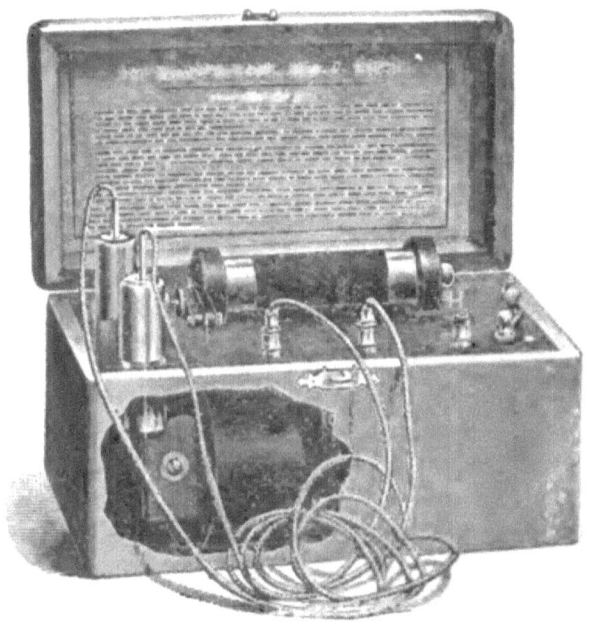

Fig. 103.—Induction Coil with Dry Cell and
Internal Damping Tube.

instance is supplied by a dry voltaic cell,
situated in the base of the apparatus. An
example of the third method is represented
in Fig. 104, where the tube T, is advanced

over the surface of the secondary coil. The action of the tube is in all cases the

Fig. 104.—Induction Coil with External Damping Tube.

same. It forms, in reality, an independent secondary coil, having a very low resistance, consisting as it does of but a single turn of conductor. When the magnetic flux from the primary coil and core, fills and empties this tube, it sets up around it, comparatively powerful induced current strengths, owing to its low resistance. Although the induced E. M. F. may be but feeble, yet with the presence of a single turn the effect of this comparatively powerful secondary current is to produce an M. M. F. and magnetic flux with such reactive power upon the primary circuit, that the resultant magnetic flux, linked with the secondary coil, is very greatly enfeebled. The further the tube is pushed into or over the coil, the greater will be its reactive influence upon the primary circuit, and the smaller will be the induced E. M. F. in the secondary coil.

The Dubois-Raymond method of varying the secondary E. M. F. consists, therefore, of varying the mutual inductive power of the primary and secondary coils, by varying their mean distance from each other. The shield method consists in leaving the primary and secondary coils at a fixed distance, but so distributing the magnetic flux through the coils, under the influence of powerful shield secondary currents, that the resultant magnetic flux through the secondary circuit is reduced.

Notwithstanding the great variety of medical induction coils, and their seeming differences of construction, electrically they invariably consist of a primary and secondary coil in mutual inductive relation to each other, and an interrupter, to vary the frequency of the primary current. The primary coil may possess a greater or less resistance and inductance, and the two coils may be so placed as to have a greater or less mutual inductance. The number of turns in the primary and secondary coils, the number of cells to be employed to operate the instrument, and the manner in which the mutual inductive influence between the coils is varied, are merely, from an electric standpoint, incidental to the main purpose for which the coil is intended; namely, to produce at the secondary terminals an E. M. F. of a given frequency, value, and wave type, when the external resistance is assigned. That is to say, if the resistance of the external circuit comprising the body of a patient, and the electrodes connected therewith, amounts, to say 2,000 ohms, then the function of the apparatus is to produce at its secondary terminals a certain effective voltage, such as would be indicated by a suit-

able voltmeter, a certain frequency, and a certain wave character of E. M. F. If the effective pressure under these circumstances at the secondary terminals be, say 6 volts, then the effective current through the external circuit, provided that the inductance of the external circuit is small; *i. e.*, that there are no coils of many turns of wire in the circuit, will by Ohm's law, be
$$\frac{6}{2,000} = \frac{3}{1,000} = 3 \text{ milliamperes.}$$

In order to compare the relative electric effectiveness of different medical induction coils, it is only necessary to measure the effective E. M. F. maintained at the terminals of the secondary, when connected through different external inductionless resistances, the frequency of the interrupted and the wave type of E. M. F. being at the same time observed. For some apparatus a high frequency, and a smooth type of wave are the desiderata, while for others, a low frequency, and a rough wave type are desired. The coil which will supply the requisite number of volts at its secondary terminals, under all variations of load, and will, at the same time, supply these frequencies and wave type with the maximum economy, simplicity, durability and convenience, will be the most effective coil, from an electric standpoint, and no structural variations in an induction coil, whether obtained by altering its magnetic circuit, its winding, or its various parts, can do anything except to alter the frequency, the wave type and the magnitude of the E. M. F. at its secondary terminals.

It might be supposed that when a medical induction coil, of the Dubois-Raymond type, has its secondary coil as far over the primary coil as it will go, that the current strength in the secondary circuit will be a maximum. This, however, is not always the case, since the M. M. F. of the current in the secondary coil may so greatly react upon the primary flux and current, as to cause the secondary to act like a metallic shield. In other words, the secondary coil may overload the primary. For this reason, it is possible that the maximum current strength in the secondary circuit may be obtained when the coil is only partly covering the primary, say one half or three quarters.

In a continuous-current circuit, since the E. M. F. is always acting, there is no difficulty in determining its value; but since, in an alternating-current circuit, the value of the E. M. F. reaches a maximum twice in every cycle, varying on each side of zero, it is more difficult to define what is the value of the E. M. F. The E. M. F. is always defined as the *effective thermal E. M. F.* or simply the *effective E. M. F.* That is to say the value of an alternating or pulsating E. M. F. is taken as being equal to that of the continuous E. M. F., which is capable of producing the same heating effect in an inductanceless resistance. Similarly, when the electric current pulsates or varies, its effective thermal value, commonly called its *effective current strength*, is equal to that strength of continuous current which would produce in a given resistance the same heating effect. If, therefore, the effective E. M. F. at the terminals of a medical induction coil be expressed or measured as five volts effective, we mean that the E. M. F. it produces, although varying between, perhaps, 20 volts and zero, produces in a simple resistance, the same amount of heat by its current in a minute or five minutes, as 5 volts of continuous-current pressure. Consequently, the statement of the effective current strength or effective E. M. F. at the terminals of an induction coil does not express the range of current or of E. M. F. in each wave, unless the shape of the wave be known.

The connections of the medical induction coil are commonly so arranged that the E. M. F. induced in either the primary or the secondary circuit can be externally employed. We have seen that on breaking, an E. M. F. of self-induction appears in the primary circuit, as well as an E. M. F. of mutual induction in the secondary circuit, but the E. M. F. of mutual induction is practically always greater than in the primary circuit owing to the greater number of turns in the secondary coil. If P_1, P_2, Fig. 105, represent the terminals of the primary coil, and S_1, S_2 the terminals of

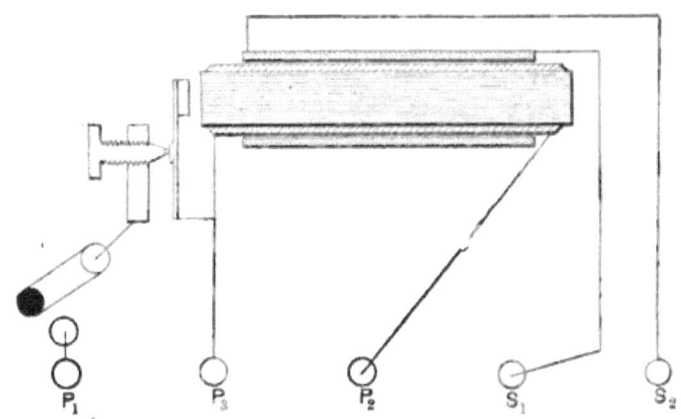

Fig. 105.—Connections of Medical Induction Coil.

the secondary coil, while P_3, is connected to the vibrating spring, then it will be seen that when the spring breaks contact, the induced E. M. F. in the primary coil will be developed between the terminals P_1 and P_2, while the mutually induced E.

M. F. will be developed between S_1, S_2. There is this difference, however, between the E. M. F. of the primary and secondary circuits, that whereas, S_1 and S_2, will supply an E. M. F. both at making and at breaking, in opposite directions, although the E. M. F. at breaking is, as we have seen, much stronger than the E. M. F. at making; yet, on the other hand, between P_2 and P_3, the E. M. F. on making is always very small, and must be less than the E. M. F. between P_1 and P_2; that is to say, it must be less than the E. M. F. of the battery. Under ordinary circumstances, therefore, when the primary coil has comparatively few turns of coarse wire, the current obtainable between P_2 and P_3, is practically zero at making, and rises suddenly and abruptly on breaking, or acts exactly in the same manner as a secondary coil of the same number of turns. Such a type of medical induction coil is represented in Fig. 106.

In Fig. 107, a common form of connection of the circuits of a medical induction

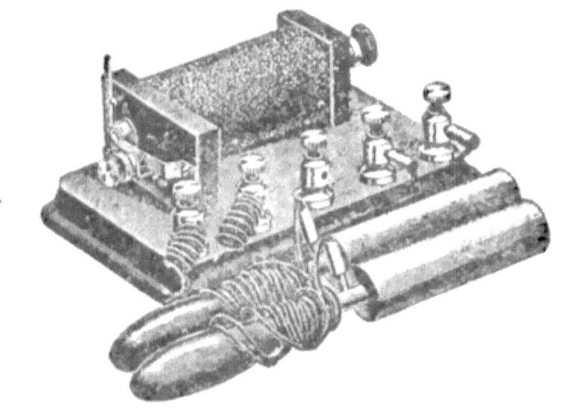

Fig 106.—Induction Coil with Handle Electrodes.

coil is shown. Here one end of the secondary coil is brought into connection with one end of the primary coil. By this means we obtain between P_2 and S_1, the self-induced E. M. F. of the primary, and between S_1 and S_2, the mutually induced E. M. F. of the secondary coil. Between P_2 and S_2, the total E. M. F. in both coils is obtained. The effect in fact is to add to the secondary coil another layer which is

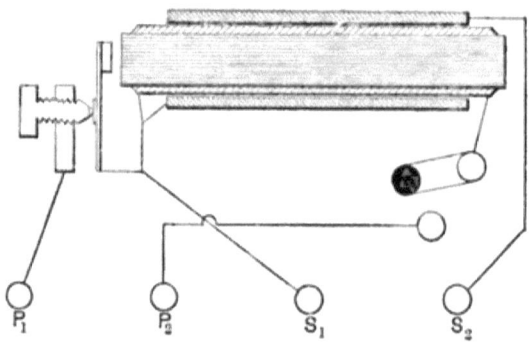

Fig. 107.—Inter-connection of Primary and Secondary Windings in Medical Induction Coil.

actually serving, before breaking contact, as a primary. At making, however, the effect due to the secondary coil is greater than the effect due to the primary coil. It is to be remembered that when any combination takes place, by the superposition of a secondary current upon a primary current, or of either upon a continuous current in a circuit, the result will be the sum of all the influences acting independently, and will consist electrically, of a pulsating or alternating wave of current, having a wave type which may be modified by the simultaneous influence of the various components.

Fig. 108, shows a form of apparatus consisting of a medical induction coil and of a battery of silver chloride cells. The induction coil can be operated from a few of the cells in this battery so as to supply dissymmetrical alternating currents of, perhaps, 200~, while the battery, which will probably have an E. M. F. of 50 volts, will be capable of supplying a continuous current.

ELECTRO-THERAPEUTICS. 295

We have already seen that both the E. M. F. and the current strength at making,

FIG. 108.—INDUCTION COIL WITH BATTERY OF CHLORIDE OF SILVER CELLS.

are smaller than the E. M. F. and current strength on breaking. It is to be remembered, however, that in all cases, the total quantity of electricity which passes through the circuit is the same in each alternating-current wave. In the secondary circuit of an induction coil, no matter how great may be the dissymmetry of wave type, that is to say, no matter if the current strength is much greater on making than on breaking, it will be correspondingly briefer in duration. In, for example, Fig. 97, which represents dissymmetrical alternating currents produced in the secondary circuit of an induction coil, the wave *abcd*, in one direction, or above the line, being produced at making contact, and *def*, the greater wave, being produced at breaking contact, then the area of the wave *abcd* as represented graphically, will be equal to the area *def*, under all circumstances. This area also represents the total quantity of electricity, measured in coulombs, which will pass through the circuit. Consequently, all physiological effects, which depend upon current strength, will be dissymmetrical or polar in the secondary circuit; that is to say, the effect produced at making will be different from the effect produced on breaking and, consequently, the effects at one pole will be different from those at the other; but all physiological effects, which depend only on the quantity of electricity, will be the same both on making and on breaking. Thus, it is well known that muscular excitability is a function of the current strength, so that the muscular excitation produced by the making and breaking currents will be different. On the other hand, all the electrolytic effects, which are dependent upon the quantity of electricity which passes, will be alternately produced in equal amounts at each wave. Consequently, either electrolysis will not appear at all, or the products of electrolysis will appear in equal quantities at each electrode.

To sum up, the medical induction coil gives discharges possessing the following characteristics:

It produces dissymmetrical alternating-current waves of fairly high frequency, but intermittent, the waves being separated by intervals of no current. The waves on breaking are steep and abrupt, except in the case of very long, fine wire coils, with large self inductance, and choking effect. The E. M. F. induced in the secondary coil may be hundreds of volts, but the E. M. F. available at the terminals, under any ordinary load, will be not usually more than 15 volts effective.

CHAPTER XI.

DYNAMOS, MOTORS AND TRANSFORMERS.

A FORM of electric source, occasionally used in electro-therapeutics, is to be found in the *dynamo-electric generator*. Of the sources already discussed, the voltaic battery produces a continuous E. M. F.; the influence machine, a pulsatory E. M. F.; and the induction coil an alternating E. M. F. Dynamos, on the contrary, can be employed to produce either continuous or alternating E. M. Fs., according to their construction. Those producing continuous E. M. Fs. are generally termed simply dynamos, or *generators;* while those which produce alternating E. M. Fs., are usually called *alternators*.

The fundamental principle employed in all dynamos and alternators, is the induction of E. M. F. by filling and emptying conducting loops with magnetic flux. Dynamos and alternators generally consist of a fixed and a rotary part, by the mutual action of which conducting loops become filled and emptied with magnetic flux during rotation. The E. M. Fs. so generated will be alternating, unless a commutator be employed to render them continuously directed in the external circuit. Any dynamo, therefore, which employs no commutator is necessarily an alternator. This has been already illustrated in the case of the hand magneto-generator described in Chapter VIII.

The E. M. F. obtained from a continuous-current generator is always slightly pulsating, as represented in Fig. 40. The departure from a continuous E. M. F., such as is produced by a voltaic battery, is less marked when the number of commutator bars is great. When, however, few bars are employed, the pulsation may be more evident, as shown in Fig. 41. If it were possible to employ an indefinitely great number of commutator bars, the E. M. F. would be unvarying. This deviation from continuity occurs at each passage of a commutator bar or segment beneath the brush.

A dynamo, whether continuous or alternating, usually employs electromagnets to supply the flux with which its loops of wire are filled and emptied. In the case of the magneto-generator illustrated in Fig. 93, a permanent magnet is employed for this purpose. Electromagnets require to be supplied with a continuous current for their excitation. This exciting current is either supplied from the armature of the dynamo itself, or from some separate source. In the former case the machine is said to be *self-excited*, and in the latter, *separately-excited*. When the external circuit of a continuous or alternating-current generator is opened, the machine requires no more power for its operation than that necessary to overcome its frictions, together with the small amount of electrical power supplied for the excitation of the field magnets. When, however, the machine furnishes a current to an external circuit, the power which has to be supplied to drive the dynamo is increased by the amount of electric power in the external circuit. If, for example, a machine expends 50 watts, or volts-amperes, in its external circuit, the power applied to drive it is increased by something more than 50 watts, since some loss of power occurs in the machine in order to furnish the external work.

Continuous-current generators are not very extensively used in electro-therapeutics, since other sources of continuous E. M. F. are available. When generators are used, they are generally driven by electric motors, taking their power from neighboring electric circuits. For example, if it be required to obtain a low continuous pressure, of say 5 volts, in order to operate an electric cautery, and an electric circuit of higher pressure, be available in the neighborhood, say of 110 or 220 volts pressure, as commonly used in electric lighting, a small motor, operated from the lighting mains, may be employed to drive a dynamo constructed to give the necessary E. M. F. and current required for the cautery knife.

Electromagnetic motors are operated either from continuous or alternating-current circuits. A motor, however, which is arranged to work on a continuous-current circuit is not usually capable of operating on an alternating-current circuit, and *vice versa*. Moreover, in either case a motor will only operate effectively at the E. M. F. for which it is designed.

A particular form of small motor, operated from 110 volt continuous-current pressure, is represented in Fig. 109. M, M, are the field magnets, A, the armature, CC, the commutator, B, one of the brushes resting upon the commutator, and P, the pulley.

Another form of small motor intended to be driven by a primary or storage battery is shown in Fig. 110. A form of small

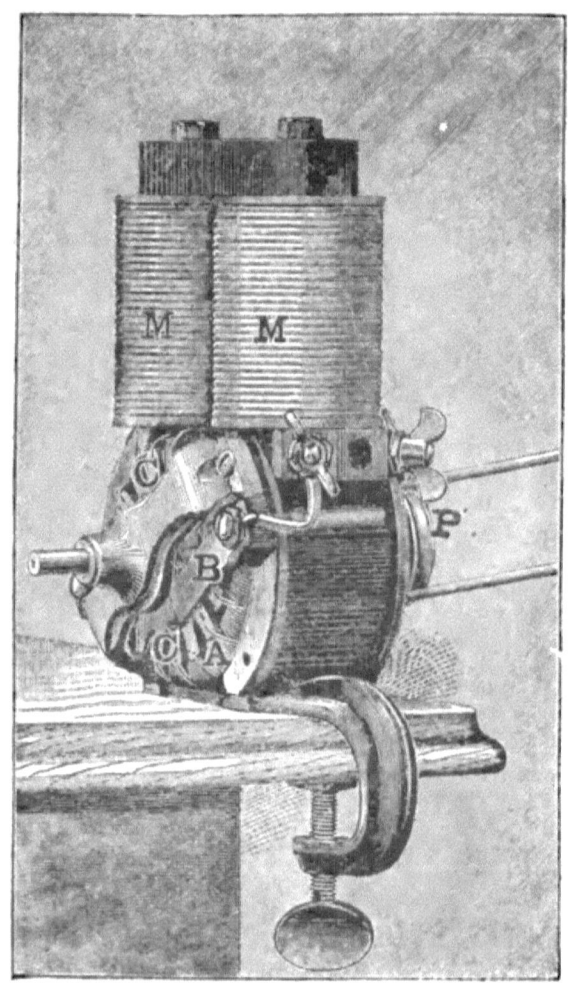

Fig. 109.—Electromagnetic Motor.

alternator designed for electro-therapeutic purposes is represented in Fig. 111. The coils *C, C, C*, on the field frame, are wound

Fig. 110.—Electromagnetic Motor.

with two circuits; a coarse wire circuit excited by a continuous current from a pair of binding posts *P*, and a fine wire circuit connected with the binding posts *S*. The armature *AA*, which is rotated by a small pulley at one end of the shaft, is constructed of sheets of soft iron, and carries teeth T, T, in such a manner as alternately to open and close the magnetic circuits of

Fig. 111.—Electro-Therapeutic Alternator.

the coils C, C, C. There are 12 poles in the field frame, so that each revolution of the armature produces 12 complete periods, or 24 alternations. As soon as the teeth bridge across adjacent poles, magnetic flux· is poured through the secondary circuits, or fine wire circuits, inducing in them an E. M. F. in one direction, and as soon as the teeth pass beyond this position, the magnetic circuits are opened, and the secondary coils are emptied of flux, thus inducing an oppositely directed E. M. F. The advantage of such an alternator is that it furnishes alternating E. M. Fs. of approximately sinusoidal type, and at a frequency which, within certain limits, is under control. At the high speed of 4,800 revolutions per minute, or 80 revolutions per second, the frequency of alternation will be $80 \times 12 = 960$ complete cycles, or 1,920 alternations per second; while at lower speeds the frequency will be correspondingly reduced. The E. M. F. obtainable from such a machine is about 50 volts, The E. M. F. at terminals is, however, considerably less than this when ordinary loads are applied.

Alternating currents are frequently supplied from electric lighting stations to consumption circuits and buildings, at a comparatively high pressure, 1,000 or 2,000 volts effective being the pressure commonly employed. As this is a dangerously high pressure to handle, it is never permitted to enter a house, the pressure being reduced at some point outside the house, by an apparatus called a *step-down transformer*, which is a form of induction coil in which the primary wire contains a greater number of turns than the secondary. Alternating currents generated by large alternators, placed in the central station, are sent through the primary coils of the transformer, usually by overhead wires. The secondary coils of the transformers generate a pressure of 50, 100 or 200 volts, according to circumstances, and wires from

FIG. 112.—ALTERNATING CURRENT TRANSFORMER.

the secondary coil enter the building to be supplied. A form of transformer is shown in Fig. 112, P, P, being the primary, and S, S, the secondary wires. The ratio of the secondary to the primary pressure is called the *ratio of transformation*. Thus, if the primary pressure between the wires P, P, be 1,000 volts effective, and the secondary pressure between the wires S, S, 50 volts effective, the ratio of transformation is 1 : 20, and this will be approximately the ratio of the number of turns in the secondary coil to the number of turns in the primary coil of the transformer. The frequency of alternation employed in alternating-current electric lighting is not higher than 140$\sim$, or 280 alternations per second, and usually varies between this and 125$\sim$ or 250 cycles. In some cases, however, the frequency may be 60$\sim$ and even as low 25$\sim$ per second.

A particular form of transformer, designed for supplying alternating electric currents at the pressure required to operate a cautery knife, is shown in Fig. 113. *PP*, is the primary coil wound upon

Fig. 113.—Alternating-Current Transformer for Cautery.

an iron core, consisting of a bundle of straight iron wires and resembling, therefore, in general form, the primary coil of a medical induction coil. The secondary coil *S*, consists of a short coil of thick wire, which, having a low resistance, enables comparatively powerful currents of say 5 to 30 amperes to be produced in the secondary circuit. The apparatus is, therefore, a step-down transformer. The primary coil is wound for an effective alternating pressure of 50 or 100 volts, according to the pressure employed in the lighting circuits of the building. In order to regulate the E. M. F. and current in the secondary circuit, the secondary coil S, is moved from or towards the centre of the primary coil P, by the screw S, after the manner of the Dubois-Raymond type of adjustment in the medical induction coil. The contact C, is so arranged that by the closing of the box, the primary circuit is opened.

In some cases, where continuous currents are supplied to a building for lighting purposes, at 110 volts pressure, it is

possible to dispense entirely with the use
of batteries for the operation of a medical
induction coil, or for the production of

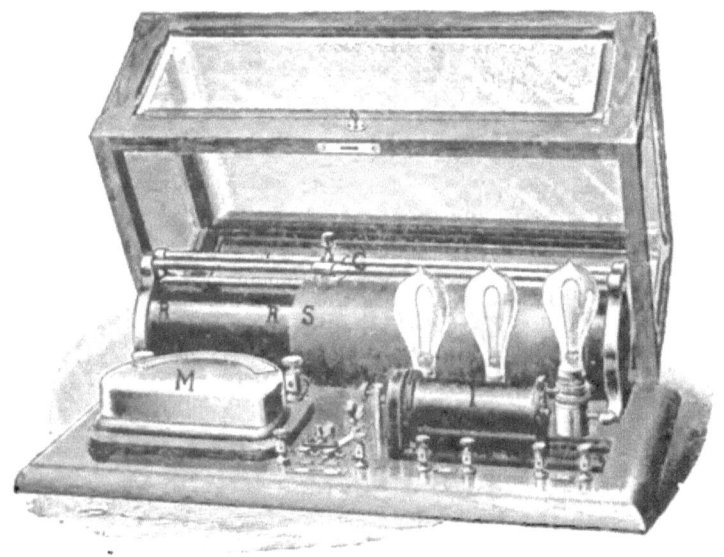

Fig. 114.—Adapter for Continuous Current Circuits.

feeble continuous currents in electro-therapeutic work. An apparatus for this purpose called an *adapter* is shown in Fig. 114. It consists essentially of a rheostat, placed in the circuit of the electric lighting mains, in such a manner as to reduce the current required to the right strength without danger. A long cylinder $RR\ SSS$, of hard rubber, contains at the end RR, a number of resistances, which are connected with brass contact strips above. The sliding contact C, makes connection with one of these brass strips, so as to include any desired number of resistances in the circuit. At the end of these resistances, and connected with them, is a long spiral of fine German-silver wire, wound in a fine groove on the cylinder, SS, so that the contact C, in sliding over the cylinder to the right, makes contact in succession with each turn of German-silver wire, thus cutting out the resistance very gradually over this portion of the circuit. M, is a milliammetre, and I, an induction coil, whose primary circuit is operated by a current from the electric lighting mains through the lamps L, L, L.

The connections of the adapter are shown in Fig. 115. It will be seen that

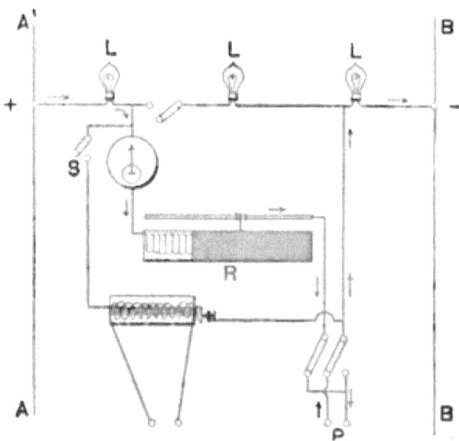

Fig. 115.—Connections of Adapter.

the current through the mains AB, passes through the lamps LL, through the circuit of the patient at P, and through the adjustable resistance and milliammetre. By connecting the middle lamp L, in the circuit, the pressure connected with the patient can be considerably reduced. It will be observed, that in no case can the circuit of the patient be connected to the mains without the interposition of two 110-volt electric lamps. The primary of the induction coil can be thrown into circuit by the use of the switch S.

The fact that an incandescent electric lamp can be entirely enclosed in a nonconducting air-tight glass chamber, renders it suitable for introduction into the cavities of the body. Two miniature incandescent lamps, suitable for such exploratory purposes, are shown in Fig. 116. These give about half a candle, and are operated at pressures of between 2 and 4 volts, with a current strength of from 1 to 1 1/2 amperes. Care must be taken in the operation of such lamps, that the pressure shall not exceed that for which they are designed, as otherwise the lamps will be destroyed. They can be supplied with either an alternating or a continuous current, but are usually operated by a battery. Since a 1/2 candle-power lamp

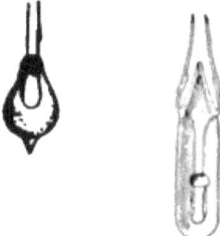

Fig. 116.—Incandescent Electric Lamps for Exploration.

requires an activity of about 3 1/2 watts, or at the rate of 7 watts per candle, while a lamp of 8 candle-power requires to be supplied with about 30 watts, or at the rate of about 3 1/2 watts per candle, it is evident, that when the lamp has considerable illuminating power, the heat it liberates may be inconveniently great. Consequently, when the candle-power of lamps for exploratory purposes exceeds a certain amount, it is customary to enclose the globe in a second glass chamber, through which water is circulated, so as to carry off the surplus heat.

The heating power of the electric current is often applied in surgery for cauterizing purposes. *Electric cautery knives* consist essentially of suitably shaped platinum wires, heated by the electric current. Fig. 117, shows several forms of such cautery knives. The amount of activity required to render the cautery knives white hot, depends upon the surface of hot platinum which they expose to the air. A broad, flat knife requires more activity than narrow blades. Either alternating or continuous currents are suitable for cautery knives. Either primary or secondary cells are frequently employed for this purpose. For the broadest knife

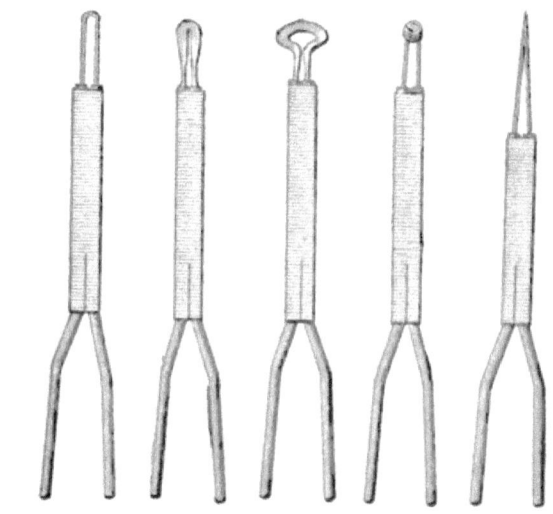

Fig. 117.—Electric Cautery Knives.

in the figure, 25 or even 30 amperes, at a pressure of approximately one volt, may be required, representing an activity in the knife of from 25 to 30 watts. In the *platinum snare cautery*, a growth or part is removed by causing a length of wire to encircle the part and then drawing the loop tight, so that the glowing wire is pulled through the part to be removed. Here, owing to the length of wire which has to be heated, though the total activity may be comparatively small, yet the E. M. F. necessary to send the required current through the length of platinum wire may be considerably greater than that for a cautery knife.

We have seen, that in accordance with Ohm's law, the current strength in any circuit may be altered, either by varying the E. M. F., or by varying the resistance. Both of these methods are employed in electro-therapeutics. Instruments for varying the resistance in a circuit are called *rheostats*. They consist essentially of resisting paths whose length or area of cross-section may be adjusted, or varied at will. In most forms of rheostat, it is the length of resisting path and not the area of cross-section, which is varied.

The form given to the resisting paths depends upon the strength of the current which has to be regulated. Currents for cauterizing, which may be as high as 20 or 25 amperes, require comparatively coarse wire coils; for, each ohm through which a current of 25 amperes passes, liberates heat at the rate of 625 watts, or nearly one H. P., and, consequently, if this one ohm consisted of fine wire of comparatively short length and, therefore, possessing a very limited radiating surface, the wire, being unable to dissipate this heat, would acquire a temperature, probably, sufficient to melt it. The comparatively feeble currents generally employed in electro-therapeutics do not require an extensive radiating surface, and the rheostats through which they pass may be composed of fine wire or of water or of carbon.

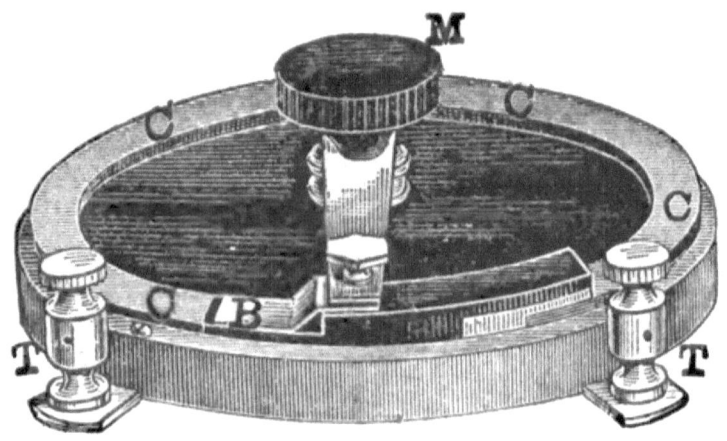

Fig. 118.—Carbon Rheostat.

One of the simplest forms of rheostat for very feeble currents is shown in Fig. 118. Here the resisting column consists of a thin layer of graphite obtained by rubbing a soft graphite pencil in a circular path around the rim of a slate slab, *CCC*. By this means a layer of fairly high resisting carbon is obtained, and the length of this path in a circuit, determines the amount of resistance included. This length is adjusted by altering the position of the brush *B*, attached to the milled-headed screw *M*. It becomes necessary in practice, to occasionally renew the carbon layer. Its resistance can be varied by rubbing more or less graphite over the surface.

Another form of carbon rheostat is shown in Fig. 119. Here the resisting path is composed of pulverized carbonaceous material pressed into a groove in an insulating plate. A number of brass studs, *CCC*, pass through the surface of the insulating plate and make contact with the carbon column in the groove beneath. The length of the carbon column inserted between the terminals, *T, T*, can be varied by turning the handle *H*, so as to make contact with the brass studs at different portions of the circumference.

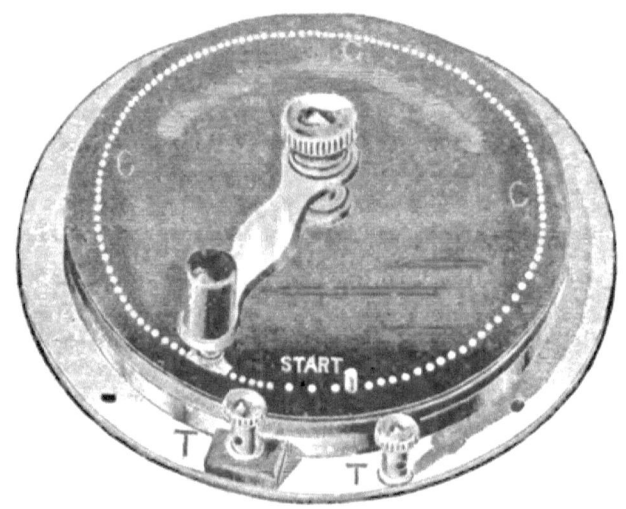

Fig. 119.—Carbon Rheostat.

Fig. 120, shows another form of carbon rheostat depending upon a somewhat different principle. Here powdered carbon is placed in a chamber provided with elastic sides *CC*. The resistance between the top and bottom surfaces of this mass of carbon depends upon the pressure which is brought to bear upon the layer. When the pressure is very light the carbon par-

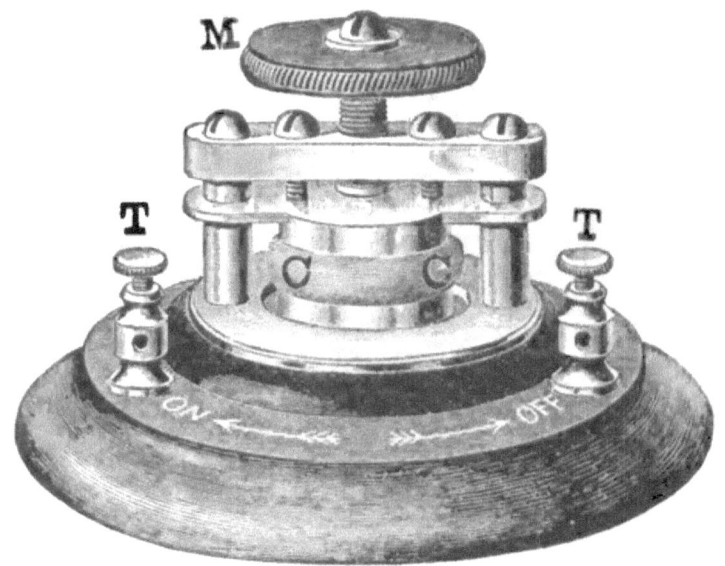

Fig. 120.—Carbon Pressure Rheostat.

ticles do not make good electric contact with each other and interpose a comparatively great resistance to the passage of the current from one to another. When, however, the pressure is considerable, the particles are brought into more intimate electric contact and the resistance of the

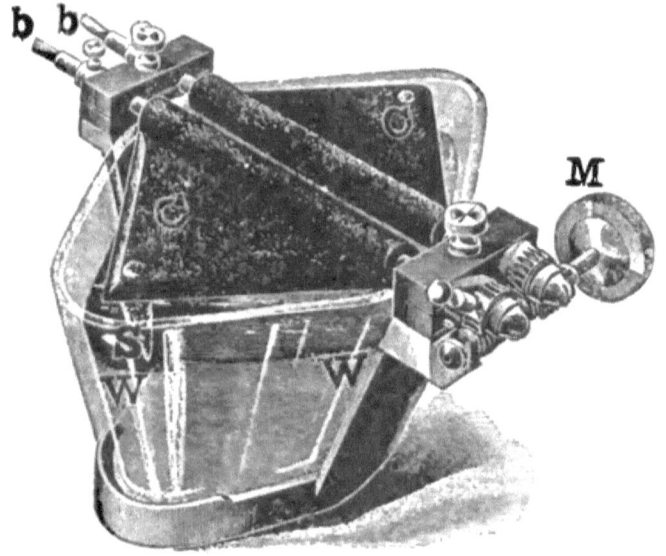

Fig. 121.—Water Rheostat.

mass is thereby greatly reduced. The pressure in this instrument is varied by turning the milled-headed screw M.

Fig. 121 represents a form of *water*

rheostat. Here the column of resisting material is composed of water, which, as we have seen, possesses a high resistivity. The binding posts bb, constituting the terminals of the instrument, are connected each to a triangular mass of carbon, CC, armed at its extremity with the small sponge S. In order to vary the resistance, the milled head, M, is turned, which by means of a worm gear rotates the carbon plates so as to move them into or out of the liquid, and thus vary both the length and cross-section of the liquid column between them.

CHAPTER XII.

HIGH FREQUENCY DISCHARGES.

ALL the electric sources we have described produce E. M. Fs., and all E. M. Fs., when permitted to do so, produce electric discharges or currents. The type and magnitude of E. M. F. determine the type and magnitude of the electric current. It is, therefore, to be remembered that, however different may be the appearance of the machine which produces an electric discharge, or however different may be the appearance of the discharge itself, the difference electrically is simply one of frequency, magnitude and wave type of E. M. F.

A high E. M. F., no matter how produced, may discharge in three ways.

(1) Convectively.
(2) Conductively.
(3) Disruptively.

Either of the two last mentioned methods may be oscillatory or non-oscillatory.

A convective discharge is the discharge which occurs in the neighborhood of points connected with a source of high electric pressure. A pressure of 20,000 volts, or upwards, will produce convective effects. Such a pressure is furnished by an electrostatic, or influence machine, so that if an upright metallic rod S, furnished with a sharp pivot point, be attached, as shown in Fig. 122, to the prime conductor of a machine, a wheel, formed of a number

ELECTRO-THERAPEUTICS. 331

of radially pointed spokes, supported on the pivot, will be set into rapid rotation by the reaction of the convective discharge of electrified air particles, that are thrown off

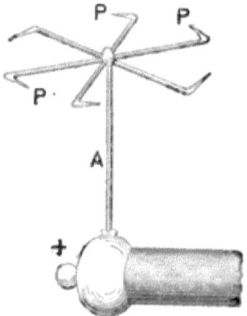

Fig. 122.—Rotation Produced by Convective Discharge.

from the points. This motion of the air produces a breeze, called a *static* or *electric breeze*, which is sometimes employed electro-therapeutically.

If a damp cord be made to connect the main terminals of a high-pressure machine, a *silent* or *conductive discharge* will pass through it, the resistance offered by such a cord being a very great number of ohms. It might be supposed, that if a metallic wire were employed instead of a string, that the discharge would pass more readily than it would through a conducting string; but, curiously enough, this is not the case, owing to the fact that the low resistance of the wire causes an enormous current strength to tend to flow through it under a high pressure at its terminals. Under the influence of this enormous rush of current, the *inductance* or *self-induction* of even a short length of straight wire, is sufficient to produce a C. E. M. F. so great, that a disruptive discharge may take place across a considerable air-gap, before any appreciable quantity can escape through the wire.

When a knuckle of the hand is approached to the rounded prime conductor of a high-pressure machine, a *disruptive discharge* or *spark* will pass through the air-gap, between the hand and the conductor. This appears to consist of a single discharge, but generally consists, in reality, of a number of separate discharges to-and-fro between the machine and the hand. In other words, the discharge is *oscillatory*, and the current oscillating. The difference between a quiet steady discharge, of a given quantity of electricity at high pressure, as compared with an oscillatory discharge of the same quantity, is shown in Fig. 123. At *A*, a steady discharge, commencing at say 100,000 volts pressure, falls steadily to zero; that at *B,* starting at an equal voltage, falls more rapidly to zero and is slightly oscillatory; that at *C,* rapidly changes direction and becomes oscillatory. The current strength in the circuit has the same graphic type in each case. The frequency of oscillation of these discharges is often exceedingly high, reaching sometimes hundreds of millions of cycles per second. The total number of

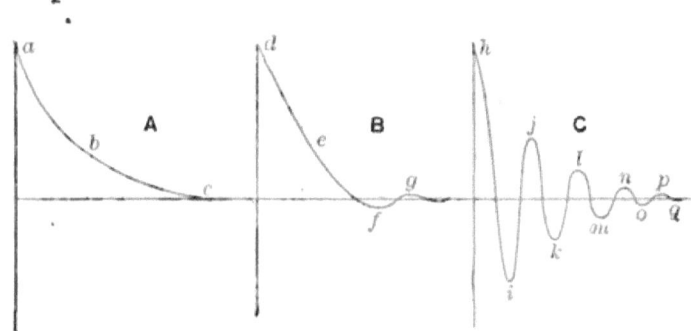

Fig. 123.—Oscillatory Discharge.

oscillations, however, in any discharge is not very great, usually varying from 2 or 3 to 20 or 30, according to the conditions of the circuit. The entire discharge, therefore, is usually completed in a small fraction of a second.

If a steel spring, such as is represented in Fig. 124, be clamped at its upper extremity, while its lower end is loaded with

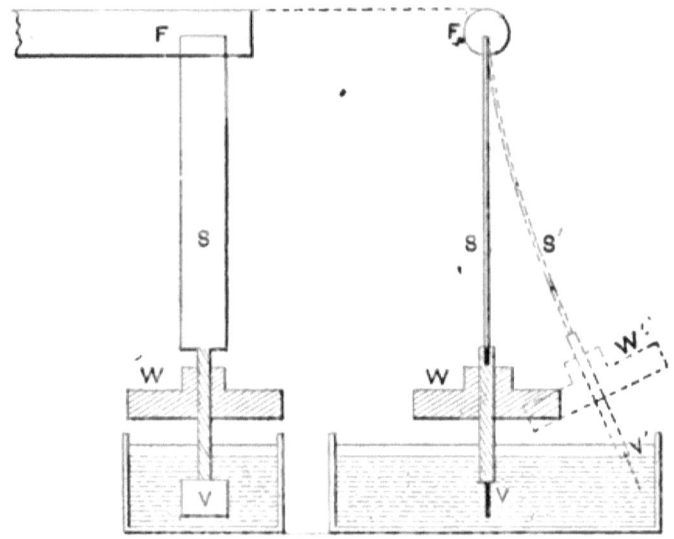

Fig. 124.—Mechanical Vibrator, Side and End View.

a weight W, and also carries a vane V, movable in a viscous liquid, then, if the spring be drawn aside from its position of rest, to the position S' W'' V', and then released, it will after a number of vibrations or oscillations return to rest, in a manner which will depend upon the frictional resistance offered by the liquid, upon the elasticity of the spring, and upon the weight with which it is loaded.

If the frictional resistance of the liquid is very great, relatively to the elasticity of the spring, such, for example, as might be offered by impulses to the motion of a large vane in molasses, then the spring will not oscillate, but will slowly return towards its position of rest. If, on the other hand, all frictional resistance could be withdrawn, not only in the vessel of liquid, but also in the air and in the molecular structure of the spring, then the spring would perform oscillations which would continue for ever, as there would then be no means for dissipating the energy of the vibrating system. In a condition intermediate between the preceding, that is to say, when the frictional resistance offered to the motion is appreciable, but not excessive, the spring will execute, by reason of its elasticity, a certain number of oscillations of successively diminishing amplitude before it comes to rest.

The frequency of the oscillations executed by the spring depends upon its elasticity, and the weight it carries. The weaker the spring; *i. e.*, the less its elastic force, the slower the vibrations; the greater the load, the slower the vibrations. In order, therefore, to produce a high frequency, we require a very stiff spring; *i. e.*, a short thick spring, and a very small weight. On the contrary, for very low frequency vibrations, we require a long and thin or weak spring, with a heavy load. Provided the frictional resistance of the liquid is not sufficiently great to check all vibration, the amount of friction will have only a very small influence upon the frequency, and will affect only the number of oscillations performed before **the system** comes to rest. In other words, if the spring oscillates, the friction **can** only damp the **system, but if the friction exceeds a certain** quantity, depending upon the size of the spring, its elasticity and load, **then** oscillation will be impossible.

Any electric circuit, in which a discharge suddenly takes place, obeys laws which **are** precisely parallel to those **we** have **above** indicated in relation to the disturbed spring. **The** frictional resistance of the liquid corresponds to the resistance of the electric circuit in ohms. The weakness of the spring corresponds to the electrostatic capacity of the circuit, or its capability of behaving as a condenser, and the load or weight added to the spring, corresponds to the inductance of the circuit; so that instead of mechanical inertia, in the electric circuit we meet *electromagnetic inertia*. When, therefore, a discharge takes place in an electric circuit, this discharge will be oscillatory or non-oscillatory, according to the amount of resistance in the circuit relative to its capacity and inductance. The greater the resistance, the less the probability of obtaining an oscillatory discharge, and when the resistance is very high, the discharge takes place slowly and without oscillation. If, however, the resistance is sufficiently small to permit oscillations or alternations of current to take place in the circuit, then the resistance will have very little effect upon the frequency of alternation, but will affect only the damping out of the vibrations. The less the resistance, the more slowly the oscillations will die out, and the greater the number that will be performed before extinction. In the same way, a circuit of large electrostatic capacity behaves like a weak spring of great length, and a circuit of small electrostatic capacity, like a small or short, stiff spring.

In order to produce very rapid oscillation or alternations in the discharge of a circuit, it is necessary to have a small condenser, and a small inductance in the circuit. On the other hand, a large condenser, discharging through a circuit of many loops of wire, and having, therefore, a large inductance, will perform slow oscillations or oscillations of low frequency. Unfortunately, however, for the production of very high frequency oscillations, a very small condenser only contains a small quantity of electricity for a given pressure applied, and, consequently, the amplitude of the current strength in the oscillations is feeble, and the total amount of energy comparatively small. On the other hand, a large jar, which will hold a large quantity of electricity, and give rise to powerful oscillations, can only produce comparatively low frequency currents. The frequency of alternating-current discharges, when of an oscillating character, and from an ordinary Leyden jar of pint size, is roughly about 15,000,000 of periods per second, when only a short length of wire is employed to connect the external and internal coatings.

In Fig. 125, a condenser or Leyden jar J is represented as being about to discharge through an air-gap in its circuit

Jabc. The discharge will be oscillatory if the resistance for the circuit be sufficiently small. Similarly, if as shown at *B*, the condenser *C*, be charged by turning the switch *S*, to *a*, the charge from the E. M. F. will be oscillatory if the resistance

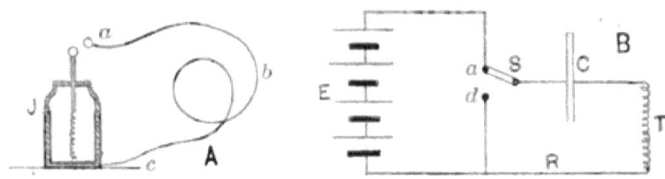

FIG. 125.—CIRCUIT FOR THE PRODUCTION OF ALTERNATING OR OSCILLATORY CURRENTS.

of the charging circuit be sufficiently low. The condenser may receive a non-oscillatory charge and give an oscillatory discharge, or *vice versa*, by properly proportioning the resistance of the circuits.

The E. M. F. of the discharge will, of **course,** be greater, the greater the distance through which the spark discharge passes, but whatever the E. M. F., the frequency of oscillation in the circuit will be the same, and only the amplitude of the waves will be affected. It will be evident, that if a very low E. M. F. be employed, the waves will be very feeble and a high E. M. F., such as supplied by an influence machine, is necessary for powerful oscillations.

The frequency of oscillation in a given circuit is not readily computed with any degree of accuracy, when the circuit is very short, owing to the fact that even a straight wire offers an amount of inductance appreciable to rapid discharges, and, although the inductance of a coil of many turns of wire can be measured or calculated, that of a short length of bent wire is difficult to estimate. The frequency of oscillation of spark discharges has been experimentally observed, in a number of cases, by observing or photographing the succession of sparks in a discharge with the aid of a rapidly rotating mirror.

It is possible, therefore, to render the charge or discharge of a circuit oscillatory by suitably regulating its capacity, inductance and resistance. Such discharges are frequently used in electro-therapeutics under the title of *static induced currents*, an unfortunately misleading term. The usual connections employed in such a circuit are shown in Fig. 126, where A and B, the main terminals of an influence machine, are maintained at a high pressure until they discharge through an intervening air-gap. Before discharge occurs, the Leyden jars j, j, become charged by this pressure, and the discharge of the jars occurs as an *impulsive discharge*, that is to say, an oscillatory discharge. The number of oscillations in the discharge will depend upon the resistance in the circuit, both of

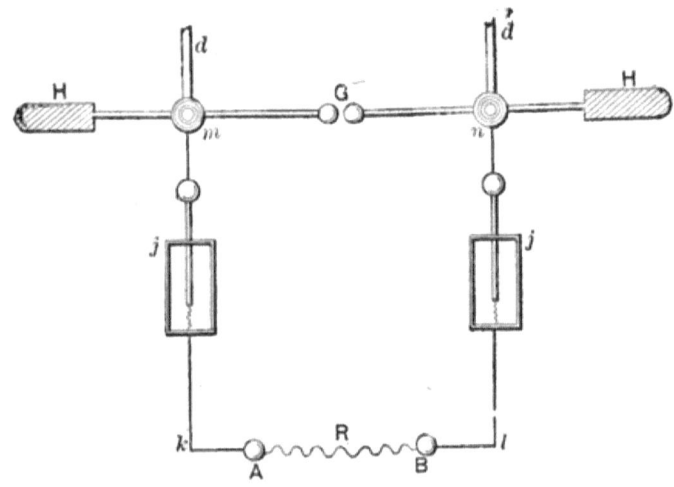

Fig. 126.—Oscillatory Current Circuit of Influence Machine.

the spark gap G, and the resistance R, of the patient between A and B. The resistance of the spark gap is not definitely known, but from the observed dampening effect in experimental circuits, its resistance appears to be only a few ohms. The frequency of oscillations depends upon the capacity of the jars and the inductance of their circuit. Under ordinary conditions, the frequency is several hundred thousand periods per second.

In order that oscillations shall be set up in the circuit CD, it is not essential that two Leyden jars should be used, although their presence assists in maintaining the insulation of the prime conductors of the machine. The spark discharge, which occurs at the air-gap G, will always be oscillatory, provided that sufficient capacity exists connected with the machine, relative to the resistance and inductance of the discharging circuit. Two equal Leyden jars, in series, are shown in the figure, connected as a single Leyden jar, and equivalent to half the capacity of either.

As already stated on page 81, the passage of one coulomb of electricity through a circuit in one second of time means a rate of flow, or current strength, of one ampere, on the average, during that time. If this coulomb passed through the circuit in one thousandth of a second, the mean current strength would be 1,000 amperes, and if in the millionth of a second, the mean current strength would be 1,000,000 amperes. For this reason, although the total quantity of electricity in a pair of Leyden jars, such as represented in Fig. 122, even when charged at a pressure of thousands of volts, is very small, yet, owing to the great frequency, or rapidity with which this charge is passed through the circuit, the current strength during that time may be very considerable. The patient placed in the circuit between A and B, may, therefore, be traversed by an alternating current of much greater strength than he would be able to receive without pain under ordinary conditions. For reasons which are yet not thoroughly understood, but which are believed to be physiological rather than physical, as soon as a certain frequency of alternation, in an alternating current is attained, the sensory effect of the current almost disappears, as though it required a certain interval of time to elapse between successive alternating currents for a nerve under the influence of this current to register sensory effects in the brain. It would seem probable, therefore, when such discharges of high frequency pass through the body, that owing to their high current strength, as well as to their high frequency, powerful physiological effects may be produced.

So far as is at present known, no current can pass through such a mass of materials as that of which the human body is composed, without effecting *electrolytic decomposition* or, in other words, that the only medium of conduction in such a mass is *chemical decomposition or electrolysis*. According to this view, rapidly alternating currents produce electrolytic effects, not only in the medium immediately surrounding the poles or electrodes, but also in the *intrapolar* or intervening tract.

It has been suggested that alternating currents of such high frequency would be unable, in traversing the human body, to penetrate more than a very moderate distance below its surface, and that, therefore, only superficial portions of the body could be directly affected by the discharge. Owing, however, to the feeble electric conductivity of the materials in the body, this *skin effect*, or tendency of the current to seek the outer layers to the exclusion of the inner layers, is comparatively small at the frequencies which can be practically produced, for, while the depth to which such currents would penetrate in good conductors, such as copper wires, is very small, yet in the case of comparatively high resisting materials, such as those constituting the human body, the penetration would probably extend practically through the entire mass. High frequency alternating currents are, therefore, powerful but painless currents, and are, probably, attended by electrolytic effects in the entire mass, although, as in the case of all alternating currents, little if any accumulation of electrolytic materials can take place.

It is not necessary to employ an influence machine for purposes of obtaining high-frequency alternating currents. Alternating-current dynamos; *i. e.*, alternators, can be employed to produce directly frequencies up to 10,000 periods per second, although such machines are expensive and troublesome to operate, requiring high speeds and special construction. Powerful induction coils, charging condensers, are also capable of producing high pressures through the discharging circuit, in the same manner as influence machines.

The general method employed for producing high-frequency alternating currents, is illustrated in Fig. 127. Here *S S*, is the secondary winding of a powerful induction coil, provided with a spark gap at *G*. *SS*, is preferably excited by a low-frequency alternating current, passing through its primary coil, say for example, from an alternating electric lighting circuit. The function of $s\ s'$, is to produce high pressures which charge the condenser

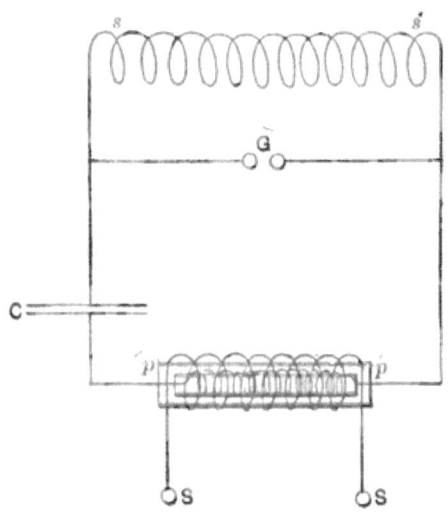

Fig. 127.—Apparatus for High Frequency Alternating Currents.

C, with a quantity of electricity proportional to this pressure. The spark gap G, is so adjusted that the pressure is able to discharge across it and in such discharge to permit the condenser C, to empty its charge through the circuit $CppGC$, with oscillations depending for their frequency upon the capacity of C, and the inductance in the circuit comprising the coil pp. This coil pp, consists of comparatively few turns of well-insulated wire, and serves as the primary winding of an induction coil, whose secondary SS, has also comparatively few turns, although more than the primary pp, but the turns are very carefully insulated from each other. The effect of passing these very rapidly alternating currents through the primary pp, is to set up, by mutual induction, very powerful induced E. M. Fs. in SS, of the same frequency, which E. M Fs. may be utilized directly. These very powerful induced E. M. Fs. are capable under suitable conditions, of giving sparks several feet in length, and these discharges, representing the surgings of hundreds of thousands of volts, are, nevertheless, almost painless, owing to their high frequency. One of the methods

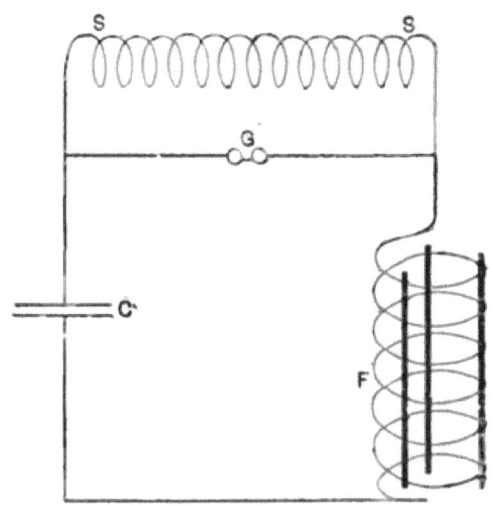

Fig. 128.—Apparatus for Creating Rapidly Oscillating Magnetic Field.

which have been applied electro-therapeutically, in connection with high-frequency currents, is represented in Fig. 128. Here the secondary winding SS, of a powerful induction coil, charges and discharges the condenser C, through a large solenoid or open coil F, of comparatively few turns, upon a vertical cylindrical frame about six feet high. The patient is introduced into this frame. His body acting as a secondary circuit, induced alternating currents circulate around it generally parallel to those in the primary coil.

When the best results are desired from high frequency discharge, it is essential that the knobs between which the spark discharges take place, shall be brightly polished. If this precaution is not taken, the discharges across the air-gap are apt to assume a convective rather than a disruptive character.

CHAPTER XIII.

ELECTROLYSIS AND CATAPHORESIS.

THE more modern theory of electrolysis regards the conduction of electric currents through all substances except metals as a convective action, in which only free atoms or radicals can take part ; that is to say, a molecule of any substance is incapable of conducting electricity, except in the case of metals. Where molecules, however, are *dissociated* into their atomic constituents ; *i. e.*, into free atoms or radicals, these constituents are capable of receiving and conveying electric charges, and so become the medium of transport in an electric current. As a consequence of this, the atoms, after having delivered their electric charges, accumulate at the electrode to which they are directed.

A molecule invariably consists of two distinct parts called *ions*, or *radicals*, named respectively the *electro-positive* and the *electro-negative ion* or *radical*. When electrolytic decomposition of the molecule occurs, it is the electro-positive radical or ion which appears at the negative electrode, called the *cathode*, or the upway, and the electro-negative radical or ion which appears at the positive electrode, called the *anode*, or the downway. When, for example, a current is led between platinum electrodes, through hydrochloric acid, it is supposed that there exists in the liquid besides the hydrochloric acid molecules proper, a considerable number of atoms or radicals of hydrogen, and of chlorine, in an uncombined state, resulting from the decomposition of some of the molecules. When an E. M. F. is connected to the electrodes these atoms receive electric charges, the hydrogen, or positively charged atoms, being attracted to the negative electrode, and the chlorine, or negatively charged atoms, to the positive electrode, so that the passage of the current is accompanied by two streams of ions moving in opposite directions through the liquid.

When a quantity of electricity passes through a liquid, the products of electrolytic decomposition, collected at the electrodes, are found to be in strict proportion to the quantity of electricity which has passed. Every coulomb of electricity, in its passage through the solution, leaves at the electrodes a definite number of ions differing in different liquids. Thus, when hydrogen is liberated, this quantity is 0.01038 milligramme-per-coulomb; so that one ampere; *i. e.*, one coulomb-per-second, passing through a solution and liberating hydrogen, will liberate 0.01038 milligramme-per-second.

A certain critical value of the E. M. F. is required between the electrodes, in a liquid, before electrolysis can take place. In other words, a liquid offers a certain C. E. M. F. having a definite minimum value, and this C. E. M. F. must be overcome, in addition to the C. E. M. F. due to ohmic resistance, which the liquid possesses by virtue of its resistivity and geometrical proportions, before the current will pass through the liquid.

When liquids that are capable of mixing, are placed in a vessel, in compartments separated from each other by porous partitions, an unequal mixing of the two takes place through the pores of the partition or septum. This unequal mixing through the pores of the intercepting medium is called *osmose*. Under its influence, the level of the liquids on opposite sides of the septum will be changed. In the case, for example, of sugar and water, placed on one side of the septum, formed of say a piece of hog's bladder, and pure water placed at the same height on the other side, the liquid current from the pure water is stronger than the current from the sugar and water, so that the level of the liquid rises on the side of the sugar and water. The two currents are sometimes distinguished as follows; viz., the one towards the higher level is called the *endosmotic* current, and the one toward the lower level, the *exosmotic* current.

When an electric current is sent through a porous septum separating either two different liquids, or two portions of the same liquid, some of the liquid is transported bodily through the septum, almost always in the direction of the electric current; that is to say, from the positive pole or anode toward the negative pole or cathode. This is called *electric osmose* or *cataphoresis*. This electric osmose takes place independently of ordinary osmose, and since its direction varies with the current, it may be made to either aid or oppose ordinary osmose. The quantity of liquid transported depends both on the nature of the liquid and on the nature of the porous diaphragm, but in every case is directly proportional to the quantity of electricity which passes. The quantity transported in a given time is, therefore, proportional to the current strength. The thickness and area of the porous diaphragm have no effect upon the amount of liquid transported, provided the current strength is constant, but it is evident, that a thick septum or diaphragm with a small active surface, will add a greater resistance to the circuit than a thin diaphragm, of large active surface, and, consequently, will tend to restrict the current strength, and, therefore, the amount of liquid transferred. With any given porous membrane the rate of transfer, or the quantity of liquid transferred per-coulomb of electricity, is directly proportional to the resistivity of the liquid; the higher the specific resistance or resistivity of the liquid, the greater will be the amount of liquid transferred.

For the above reason, a very dilute solution of a salt in water is much more rapidly transferred through a porous diaphragm by electric osmose or cataphoresis, than a dense or nearly saturated solution, since the dilute solution has a greater resistivity, or smaller conducting power, but the total quantity of salt transferred in a given mass of dilute solution will be less than that transported in the same quantity of concentrated solution, so that the advantage of employing a dilute and rapidly transported solution frequently disappears.

Since the human skin, from a physical point of view, is a porous diaphragm, it is possible to cause almost any solution to be transferred through it into the subjacent tissues, by placing an electrode thoroughly moistened with the solution over the portion of the skin selected, and connecting it with the positive terminal of the source of continuous E. M. F. employed, while the negative electrode is placed in contact with some other portion of the body. Methods of treatment based upon this action, whereby drugs or medicaments are directly introducted into the parts to be acted on, are called *cataphoric medication*.

The combination of electrolysis with cataphoric medication is sometimes called *metallic electrolysis*. Thus, if a moistened copper electrode be placed over a surface of skin, or mucous membrane, and be connected with the positive pole of a source of continuous E. M. F., while the negative pole is placed in connection with some other part of the body, a salt of the metal will be formed at the surface of the metal by electrolysis, and this salt, entering into solution at the surface of the skin, will be carried through the skin or membrane by cataphoric action.

CHAPTER XIV.

DANGERS IN THE THERAPEUTIC USE OF ELECTRICITY.

An uninsulated electric conductor carrying a current, though held in the hand, is not dangerous from the passage of the current, unless the wire is so overheated that the wire becomes dangerously hot. If, however, a high E. M. F. be connected with the wire, then holding such wire in the hand may become dangerous, if a circuit be established through the body for the passage of an electric current from the E. M. F. That is to say, the body may receive a dangerously powerful electric current. A man standing on a dry, wooden floor, may safely touch a wire connected with a high pressure. For example, he may hold in his hand a bare wire leading from a dynamo supplying a number of arc lamps in series, and, therefore, having a difference of electric potential, relatively to the ground, of several thousand volts. The man will, probably, be absolutely unconscious of any effect produced by the current passing through the wire. But, should the man, while touching this wire, come into contact with some other electric conductor, such, for example, as an iron beam connected with the ground, or a grounded wire, he may receive a dangerously **powerful, and even** fatal, electric current through his body; for, if a ground exists anywhere in the circuit of the wire he holds, he will thus permit an electric circuit to be closed through his body, having in it a considerable E. M. F.

In other words, merely touching at one point a circuit through which a powerful current is passing, is not sufficient to cause a current to pass through the body. Not only a point of entrance, but also a point of exit and a complete circuit must be provided through the body, before a current can be received. It is for this reason, that the rule is frequently adopted in electric lighting stations, where conductors carrying high pressure currents are employed, always to keep one hand in the pocket when touching a conductor. By this means, if the floor is insulated, it will be very difficult to establish a circuit through the body.

The current strength which will be received by the body under any given conditions in which a circuit is established will, of course, depend upon the E. M. F. and on the resistance of the entire circuit in which the body is introduced, according to Ohm's law. The resistance of the human body may vary enormously, as already pointed out, so that it is almost impossible to say what the current strength will be in any particular case, but, generally speaking, the greater the surface area of skin coming into contact with the electrodes, and the moister the skin, the greater will be the danger of receiving a fatal shock from a powerful E. M. F.

Generally speaking, a continuous E. M. F. of 20 volts, applied anywhere to the human body through the unbroken surface of the skin, may be regarded as harmless, since the current strength that can be made to pass through any portion of the body by means of such an E. M. F. is very feeble. Alternating E. M. Fs., at frequencies commercially employed, may be painful under certain circumstances at pressures as low as even 5 volts; as, for example, when the hands are immersed in a jar of saline solution, and these jars are connected with an alternating pressure of 5 volts effective. As the pressure is increased above 5 volts of alternating E. M. F., or 20 volts of continuous E. M. F., the physiological effects become more painful, and the continuance of such a current may produce serious effects. Fifty volts of alternating E. M. F. is capable of killing a dog, in two or three seconds, when suitably applied through large wet electrodes, in such a manner as to meet with a comparatively reduced resistance in the body of the animal.

At ordinary commercial frequencies, it would appear, from experiments conducted upon dogs, horses and cows, that the danger of a given alternating-current pressure is two to three times as great as **that of** the same amount of continuous-current **pressure,** and, moreover, under the action of a powerful alternating current, the animal is deprived of volitional control of its muscles, which are thrown **into** tetanic rigidity, a much greater **strength of** the continuous current being necessary **to** produce a similar effect, even in a partial degree. At extremely high frequencies, however, far above **those at** present **commercially** employed, we have seen that the physiological effect of alternating currents is considerably less than that of continuous currents of the same strength.

Under ordinary circumstances, a man receives a shock from a wire through his hands and feet. A pressure of 100 volts continuous is not much more than appreciable when the hands are dry, and the same may be said of 50 volts of alternating current. A pressure of 500 volts is capable of giving a very severe shock, especially when a man standing on the wet ground, touches a conductor in connection with a trolley wire, at about 500 volts pressure. Rare instances are said to have occurred in which this continuous current pressure has been fatal to man. Such a pressure is very readily capable of killing a horse, partly owing to the fact that its skin is almost entirely unprotected. It would also appear from such experimental knowledge as we possess that animals are more readily killed by electric pressures than human beings.

The current strength which it is dangerous to employ depends both upon its point of application and upon its duration. A current of 250 milliamperes is, in some cases, harmless when conducted through portions of the human body for a short interval of time, while, in other cases, this strength of current, passed through vital organs, might produce fatal results. In continuous-current strength, however, any excess of 25 milliamperes is usually attended with pain under normal conditions, and is, therefore, regarded as a strength of current that should only be administered with due precautions. Even this strength of current through delicate organs, such as the eye, might produce serious results.

INDEX.

A

Action of Electrified Sphere, Mechanical Model of, 149, 150.
Active Conductor, Magnetic Flux Paths of, 192, 193.
―― Loop, Influence of, on Magnetic Needle, 193, 194.
Activity, Definition, 125.
――, Electric, Unit of, 126.
――, Mechanical, Unit of, 125, 126.
Adapter Connections for Continuous-Current Circuits, 316, 317.
―― for Continuous-Current Circuits, 314, 315.
Aero-Ferric Magnetic Circuit, 196.
Alternating Current, 121.
―― Current Dynamo, 119.
―― Current Magneto-Electric Generator, 246, 247.

Alternating-Current Transformer, 309, 310.
——— Current Transformer for Cautery, 312, 313.
——— E. M. F., 113, 114.
Alternation, Definition of, 117.
Alternator, Electro-Therapeutic, 307, 308.
Alternators, 199, 299.
Amalgam for Frictional Electric Machines, 141.
Ammeter, Definition of, 90.
Ampere, 81.
———, Definition of, 90.
Ampere-Turn, Definition of, 207.
Animal Electricity, Conclusions in Regard to, 9, 10.
Anode, 357.
Apparatus for High-Frequency Alternating Currents, 352.
Armature of Electromagnet, 204.

B

Battery, Chloride Storage, 57.
——— of Silver Chloride Cells, 39.
———, Voltaic, Definition of, 50.
———, Voltaic Plunge, 52.
Begohm, Definition of, 66.
Bichromate Voltaic Cell, 41, 42.
Bluestone or Gravity Voltaic Cell, 29, 30.
Body, Human, Electric Resistance of, 76, 77, 78.

Body, Human, Electrolytic Decomposition Produced in, 349.
———, Human, Heat Produced in by **Different** Current Strengths, 135, 136, 137.
Bonetti Electrostatic Machine, 180.
Breeze, Electric, 331.
———, Static, 331.

C

C. E. M. F., Produced by Chemical **Decomposition**, 133.
———, Produced by Magnetic Activity, 133.
———, Produced by Resistance of Circuit, 132.
Calculation of Resistance, 69, 70.
Calorie, 19.
———, Lesser, 135.
Calorimeter, 133, 134.
Carbon Pressure Rheostat, 326.
——— Rheostat, 323, 324.
Cataphoresis, 361.
——— and Electrolysis, 356 to 364.
Cataphoretic Medication, 364.
Cautery, Alternating-Current Transformer for, 312, 313.
———, Electric, Knives for, 319, 320, 321.
———, Platinum Snare, 320, 321.
Cell, Charged, 54.
———, Exhausted or Run Down, 54.

Cell, Primary, Definition of, 54.
———, Secondary, Definition of, **54.**
———, Storage, Chloride, 56, 57.
———, Storage, Definition of, 54.
———, Voltaic, Bluestone or Gravity, 29, 30.
———, **Voltaic** Dry, 48, 49.
———, **Voltaic**, Exciting Liquid of, 30.
———, Voltaic, Leclanché, 27, **28.**
———, Voltaic, Partz Gravity, 46, **47.**
———, Voltaic, Silver Chloride Form of, 36, **37, 38.**
Cells, Voltaic, Double-Fluid, 30.
———, Voltaic, Single-Fluid, 30.
———, Voltaic, Various Couplings of, 88, 89.
Charged Cell, **54.**
Charging Current, 55.
Chemical Decomposition or Electrolysis Produced in Human Body, 349.
Chloride Storage Battery, 57.
——— Storage Cell, 56, 57.
Circuit, Aero-Ferric, 196.
———, Closed, Definition of, 34.
———, Electric, 333.
———, Electrostatic, 156.
———, Ferric Magnetic, 196.
———, Magnetic, 191.
———, Magnetic, Character and Dimensions, Effect of Reluctance of, 209, 210.

Circuit, Magnetic Methods of Varying M. M. F. of, 212.
—————, Non-Ferric Magnetic, 196.
————— of Alternating-Current Transformer, 199.
Classification of Electric Sources, 26.
Closed Circuit, Definition of, 34.
Coil, Faradic, 248.
—————, Inducing, 233.
—————, Induction, Simple Form of, 249.
—————, Primary, 234.
—————, Secondary, 234.
Coils, Faradic, Adjustable Vibrator for, 274.
Comb of Points of Frictional Electric Machine, 141.
Commutator, Two-Part, Diagram of, 244, 245.
Condenser, Definition of, 175, 176.
Connections for Adapter, 316, 317.
————— of Medical Induction Coil, 290, 291.
Contact Theory, Volta's, 6, 7.
Continuous Current, 121.
————— Circuits, Adapter for, 314, 315.
————— Dynamo, 107.
————— Generators, 303.
Continuous E. M. F., 107.
Convective Discharge, 330, 331, 332.
————— Discharge, Rotation Produced by, 321.
Convention as to Direction of Magnetic Flux, 191.

Core of Medical Induction Coil, 267.
Coulomb, 80.
———, Micro, 147.
——— per second, 81.
Counter E. M. F., 130.
Couple, Voltaic, Definition of, 30.
Couplings, Various, of Voltaic Cells, 88, 89.
Current, Alternating, 121.
———, Charging, 55.
———, Continuous, 121.
———, Direct, 122.
———, Electric, 80 to 106.
———, Electric, Definition of, 80.
———, Electrostatic, 156.
———, Endosmotic, 360.
———, Exosmotic, 360.
———, Pulsating, 121.
——— Strength, Effective, 288.
——— Strength Employed in Electrocutions, 82.
Currents, Static-Induced, 344.
Cycle, Definition of, 117.

D

D'Arsonval Galvanometer, 105, 106.
Dangers in the Use of Electricity, 365 to 372.

INDEX.   379

Decomposition, Chemical, or Electrolysis Produced in Human Body, 349.
―――, Electrolytic, 83, 349.
Dielectric Medium, 157.
――― Resistance, 167, 168.
Direct Current, 122.
Direction of Induced E. M. F., Rule for, 225.
Discharge, Conductive, 331, 332.
―――, Convective, 330, 331.
―――, Convective, Rotation Produced by, 321.
―――, Disruptive, 333.
―――, Impulsive, 344, 345.
―――, Oscillatory, 333.
―――, Silent, 331, 332.
Discharges, High-Frequency, 329 to 355.
――― of Medical Induction Coils, Characteristics of, 298.
Displacement, Electric, 148.
――― Lines, 157.
Disruptive Discharge, 333.
Dissociation, Molecular, 356.
Dissymmetrical Alternating E. M. F., 119.
――― E. M. F., 118.
Double-Fluid Voltaic Cells, 30.
Dr. Ohm, 64.
Dry Voltaic Cell, 48, 49.
――― Voltaic Cell, E. M. F. of, 49.

Dubois-Raymond Type of Medical Induction Coil, 261, 262.
Dynamo, Continuous-Current, 107.
Dynamo-Electric Generator, 299.
Dynamos, 299.
―――, Alternating-Current, 119.
――― and Alternators, Fundamental Principle Involved in Production of E. M. F. by, 300, 301.
―――, Motors and Transformers, 299 to 328.
―――, Self-Exciting, 302.
―――, Separately-Excited, 302.

E

E. M. F., 25.
―――, Alternating, 113, 114.
――― and not Electricity Produced by Electric Sources, 24, 25.
―――, Continuous, 107, 121.
―――, Continuous-Current Dynamo, 110.
―――, Continuous, Graphic Representation of, 107.
―――, Counter, 130.
―――, Dissymmetrical, 118.
―――, Dissymmetrical Alternating, 119.
―――, Effective, 288.
―――, Effective Thermal, 288.

INDEX. 381

E. M. F. in Dynamos and Alternators, Fundamental Principle Involved in Production of, 300, 301.
———, Induction of, by Magnetic Flux, 221.
———, Intermittent, 112.
———, Methods of Discharge of, 329, 330.
———, Negative, Graphic Representation of, 109.
——— of Continuous-Current Dynamo, Graphic Representation of, 110.
——— of Dry Cell, 49.
——— of Edison-Lalande Cell, 44.
——— of Induction Coil, Methods of Varying Value of, 250, 251.
——— of Partz Gravity Cell, 47.
——— of Self-Induction, Direction of, on Breaking Circuit, 229.
——— of Self-Induction, Direction of, on Completing Circuit, 229.
——— of Silver-Chloride Cell, 38.
——— of Zinc-Carbon Cell, 41.
———, Positive, Graphic Representation of, 107.
——— Produced by Friction, 138, 139.
——— Produced by Friction, High Value of, 139, 140.
———, Pulsatory, 110.
———, Sinusoidal, 121.
———, Symmetrical, 118.

E. M. F., Symmetrical, Wave of, 119.
———, Unit of, 28.
E. M. Fs., Franklinic, 144.
Edison-Lalande Cell, E. M. F. of, 44.
——— Voltaic Cell, 43, 44, 45.
Effect, Skin, 350.
Effective Current Strength, 288.
——— Thermal E. M. F., 288.
Electric Activity, Source of, 128, 129.
——— Breeze, 331.
——— Calorimeter, 133, 134.
——— Circuit, 333.
——— Current, 80 to 106.
——— Current, Definition of, 80.
——— Displacement, 148.
——— Osmose, 361.
——— Resistance, 63 to 79.
——— Resistance of Flesh, 75.
——— Resistance of Human Body, 76, 77, 78.
——— Sources, Classification of, 26.
——— Unit of Work, 125.
Electricity and Magnetism, Relation Between, 184, 185.
——— and Magnetism, Transmission of, Through Vacua, 20, 21, 22.
———, Animal, Conclusions in Regard to, 9, 10.
———, Decomposition by, 83.

INDEX. 383

Electricity, Nature of, 13, 14.
———, Unit of Quantity of, 80.
Electrocutions, Current Strengths Employed in, 82.
Electrode, Negative, 357.
———, Positive, 357.
Electrodes, 357.
Electrolysis and Cataphoresis, 356 to 364.
———, Definition of, 83.
———, Metallic, 364.
Electrolyte, Definition of, 30.
Electrolytic Decomposition, 83, 349.
Electromagnet, 200, 201.
———, Aero-Ferric Circuit of, 200, 201.
———, Horse-Shoe, 202, 203.
———, Yoke of, 204.
Electromagnetic Induction, 237, 238.
——— Inrush, 338, 339.
——— Motors, 304, 307.
Electromotive Force, 13 to 62.
——— Force, Abbreviation of, 25.
——— Force, Nature of, 24, 25.
——— Force, Varieties of, 107 to 123.
Electro-Negative Ions, 357.
Electrophorus, Description of, 166.
———, Operation of, 167 to 171.
Electropoion Fluid, 41.

Electro-Positive Ions, 357.
Electrostatic Attraction and Repulsion, General Laws of, 164, 165, **166**.
―――― Circuit, **156**.
―――― Circuit, Application of Ohm's Law to, 156.
―――― Circuits of Toepler-Holtz Machine, 173.
―――― Current, **156**.
―――― Flux, **148**.
―――― Flux, Line or Curves of, **148**.
―――― Flux Paths, Representation of, **160**.
―――― Induction, **144, 145, 146,** 159.
―――― Law, General, of Attraction and Repulsion, 164, 165, **166**.
―――― Resistance, **156**.
Electro-Therapeutic Alternator, 307, 308.
Electro-Therapeutics, Galvani's Contribution to, 9.
Elements, Voltaic, **31**.
―――, Voltaic, Varieties of, **33**.
Endosmotic Current, **360**.
Ether, Luminiferous, 19, 20.
―――, Transmission of Heat by, 16, 17, 18.
―――, Universal, 14.
Exciting Liquid of Voltaic Cell, 30.
Exhausted or Run Down Cell, 54.
Exosmotic Current, 360.
External Damping Tube for Induction Coil, 281, 282.

F

Faradic Coil, 248.
—— Coil, Adjustable Vibrator for, 274, 275.
Ferric Magnetic Circuit, 196.
Flesh, Electric Resistance of, 75.
Flow, Electric, Unit of Rate of, 81.
Fluid, Electropoion, 41.
Flux Density, 213.
——, Electrostatic, 148.
——, Magnetic, 190.
——, Magnetic, Apparent Failure to Produce Physiological Effects on Human Body, 214, to 218.
——, Magnetic, Induction of E. M. F. by, 221 to 247.
——, Passage of through Human Body, 218, 219, 220.
—— Paths, Effect of Shape of Body on Directions of, 151 to 155.
—— Paths, Electrostatic Representation of, 160.
——, Remanent, 202.
——, Residual, 201.
Foot-pound, Definition of, 124.
—— per second, Definition of, 125.
Force, Electromotive, 13 to 62.
——, Electromotive, Abbreviation of, 25.
——, Electromotive, Nature of, 24, 25.

Force, Electromotive, Varieties of, 107 to 123.
———, Magneto-motive, 206.
———, Magneto-motive, Unit of, 207.
Franklin, 144.
Franklinic E. M. Fs., 144.
Frequency, Definition of, 118.
——— of Oscillation, 337, 338.
Friction, Development of E. M. F. by, 138, 139.
Frictional and Influence Machines, 138 to 184.
——— Electric Machine, Comb of Points of, 141.
——— Electric Machine, Plate Form of, 141, 142.
——— Electric Machines, 140, 141.
——— Electric Machines, Amalgam for, 141.
——— Electric Machines, Rubber of, 141.
Frog, Galvanoscopic, 2.

G

Galvani, Discovery of, 1 to 5.
Galvanometer, D'Arsonval, 105, 106.
———, Mirror, 99, 100.
———, Mirror, Sensitive, 103, 104, 105.
Galvanoscopic Frog, 2.
Gauss, Definition of, 213.
Generator, Alternating Magneto-Electric, 246, 247.
———, Dynamo-Electric, 299.
———, Magneto-Electric, 239.
Generators, 299.

Generators, Continuous-Current, 303.
Gilbert, Definition of, 207.
Graphic Representation of Continuous E. M. F., 107.
——— Representation of Oscillatory Discharge, 334.
Gravity or Bluestone Voltaic Cell, 29, 30.
Grenet's Voltaic Cell, 41, 42.
Grid of Storage Cell, 56.

H

Heat, Transmission of, by Ether, 16, 17, 18.
High-Frequency Alternating-Currents, Apparatus for, 352.
——— Discharges, 329 to 355.
——— Discharges, Physiological Effects of, 347, 348.
——— Electric Oscillations, Conditions Requisite for, 340, 341.
Holtz Influence Machine, Form of, 179.
Horse-Shoe Electromagnet, 202, 203.
Human Body, Electric Resistance of, 76, 77, 78.
——— Body, Electrolytic Decomposition Produced in, 349.
——— Body, Heat Produced in, by Different Current Strengths, 135, 136, 137.
——— Body, Passage of Flux through, 218, 219, 220.

I

Impulsive Discharge, 344, 345.
Impurities, Effect of, on Resistivity, 71, 72.
Incandescent Lamps for Exploratory Purposes, 317.
Induced E. M. F., Direction of, 225.
Inducing Coil, 233.
Inductance, 332.
——— of Secondary of Induction Coil, 264.
Induction Coil, External Damping Tube for, 281, 282.
——— Coil, Medical, 248 to 298.
——— Coil, Rapid Interrupter for, 278, 279.
——— Coil, Ribbon Vibrator for, 276, 277.
——— Coil, Simple Form of, 249.
——— Coil, Internal Damping **Tube For, 281,** 282.
——— Coils, Medical, Relative Effectiveness of, **285,** 286.
———, Electromagnetic, 237, 238.
———, Electrostatic, 144, 145, 146, 159.
———, Magneto-Electric, 238, 239.
———, Mutual, 232 to 235.
——— of E. M. F. by Magnetic Flux, 221 to 247.
——— of E. M. F. by Magnetic Flux, Varieties of, 221.

INDEX.

Induction of E. M. F., Mechanical Analogue of, 226, 227, 228.
Influence Machine, 144.
—— Machine, a form of Electrophorus, 169, 170.
—— Machine, Oscillatory-Current Circuit of, 345, 346.
Inrush, Electromagnetic, 338, 339.
Insulators, 68.
Intensity, Magnetic, Unit of, 213.
Interconnection of Primary and Secondary Windings of Medical Induction Coil, 293, 294.
Intermittent E. M. F., 112.
Internal Damping Tube for Induction Coil, 280.
Ions or Radicals, 357.

J

Jar, Leyden, 176, 177.
Joint Resistance, 73.
Joule, Definition of, 125, 126.
—— per second, Definition of, 126.
Julien Storage Battery, 59.

K

Kathode, 357.
Knives for Electric Cautery, 319, 320, 321.

L

Lamps, Incandescent, for Exploratory Purposes, 317.
Law, General, of Electrostatic Attraction and Repulsion, 164, 165.
———, Ohm's, 84 to 90.
Leclanché Cell, E. M. F. of, 28.
——— Voltaic Cell, 27, 28.
Lesser Calorie, 135.
Leyden Jar, 176, 177.
——— Jar Discharge, Oscillatory Character of, 341.
Light, Nature of, 23, 24.
———, Transmission of, by Luminiferous Ether. 19, 20.
Lines, Displacement, 157.
——— or Curves of Electrostatic Flux, 148.
Luminiferous Ether, 19, 20.

M

M. M. F., 206.
——— of Circuit, Methods of Varying Value of, 212.
Machine, Frictional Electric, 140, 141.
Machines, Frictional and Influence, 138, 183.
Magnet, North Pole of, 191.
———, South Pole of, 192.

Magnetic Circuit, 191, 192.
——— Circuit, Ohm's Law applied to, 207.
——— Circuit, Varieties of, 196.
——— Field, Rapidly Oscillating, Apparatus for Producing, 354, 355.
——— Flux, 190.
——— Flux, Apparent Failure to Produce Physiological Effects on Human Body, 214 to 218.
——— Flux, Convention as to Direction of, 191.
——— Flux, Induction of E. M. F. by, 221 to 247.
——— Flux Paths of Active Conductor, 192, 193.
——— Flux, Unit of, 211.
——— Intensity, Unit of, 213.
——— Needle, Influence of Active Loop on, 193, 194.
——— Reluctance, 206.
——— Resistance, 206.
Magnetism, 184 to 220.
——— and Electricity, Relation Between, 184, 185.
——— and Electricity, Transmission of, through Vacua, 20, 21, 22.
———, Definition of, 184.
——— Method of Producing, 188, 189.
———, Permanent, 201.

Magnetism, **Residual**, 202.
Magneto-Electric Generator, 239.
——— Generator changes in Magnetic Circuit of, 241, 242.
——— **Induction, 238, 239.**
Magneto-Motive Force, 206.
——— **Force, Unit of, 207.**
Mechanical Analogue of **Induction of E. M. F.,** 226, 227, 228.
——— Analogue of Relation Between Electricity and Magnetism, 185 to 188.
——— **Model of Action** of Electrified Sphere, 149, 150.
——— Vibrator, 335, 336.
Medical **Induction** Coil, 248, 298.
——— **Induction** Coil, Characteristics **of Discharge Produced** by, 298.
——— **Induction Coil,** Connection of Vibrator in, 269 to 275.
——— **Induction Coil,** Connections of, 290, 291.
——— Induction Coil, Core of, 267.
——— Induction **Coil, Diagram of** Primary Induced E. M. Fs., 259.
——— Induction **Coil Discharges,** Characteristics of, 298.
——— Induction Coil, Dubois-Raymond Type, 261, 262.

INDEX. 393

Medical Induction Coil, Effect of Increasing Frequency of Vibration,
—— Induction Coil, Interconnection of Primary and Secondary Windings of, 293, 294.
—— Induction Coil, Methods of Increasing Frequency of Flux Oscillations Produced by, 252.
—— Induction Coil, Methods of Increasing Magnetic Flux of, 252.
—— Induction Coil, Operation of, 254, 258.
—— Induction Coil, Primary Connections of, 253, 254.
—— Induction Coils, Relative Effectiveness of, 285, 286.
Medication, Cataphoretic, 361.
Medium, Dielectric, 157.
Megohm, Definition of, 66.
Metallic Electrolysis, 364.
Milliameter, Construction of, 94 to 98.
———, Definition of, 90.
———, Varieties of, 91 to 94.
Milliampere, Definition of, 82.
Mirror, Galvanometer, 99 to 102.
———, Galvanometer, Sensitive, 103, 104, 105.
Molecular Dissociation, 356.
Motors, Dynamos and Transformers, 299 to 328.

Motors, Electromagnetic, 304 to 307.
Mutual Induction, 232 to 235.

N

Nature of Electricity, 13, 14.
Negative E. M. F., Graphic Representation of, 109.
——— Electrode, 357.
——— Plate of Voltaic Cell, 31.
——— Pole of Voltaic Cell, 34.
Non-Ferric Magnetic Circuit, 196.
Non-Polarizable Voltaic Cells, 32.
North Pole of Magnet, 191.

O

Oersted, Definition of, 211.
Ohm, Definition of, 65, 66.
Ohm, Dr., 64.
Ohm's Law, 84 to 90.
——— Law, Application of to Electrostatic Circuit, 156.
——— Law, Application of to Magnetic Circuit, 207.
Oscillations, Frequency of, 337, 338.
———, High-Frequency, Electric Conditions Requisite for, 340, 341.
Oscillatory Character of Leyden Jar Discharge, 341.

INDEX. 395

Oscillatory Current Circuit of Influence Machine, 345, 346.
—————— Discharge, 333.
—————— Discharge, Graphic Representation of, 334.
Osmose, 360.
——————, Electric, 361.

P

Pair, Voltaic, Definition of, 30.
Parallel-Connected Resistances, 73.
Partz Gravity Voltaic Cell, 46, 47.
Period, Definition of, 117.
Permanent Magnetism, 201.
Physiological Effects of High-Frequency Discharges, 347, 348.
Plate Form of Frictional Electric Machine, 141.
Platinum Snare Cautery, 320, 321.
Plunge Battery, Voltaic, 52.
Polarization of Voltaic Cell, 32.
Pole, Negative, of Voltaic Cell, 34.
——————, Positive, of Voltaic Cell, 34.
Portable Silver-Chloride Battery, 53.
Positive E. M. F., Graphic Representation of, 109.
—————— Electrode, 357.
—————— Plate of Voltaic Cell, 31.
Primary Cell, Definition of, 54.

Pulsating Current, 121.
Pulsatory E. M. F., 110.

R

Radicals, Electro-Negative, 357.
———, Electro-Positive, 357.
——— or Ions, 357.
Rapid Interrupter for Induction Coil, 278, 279.
Rapidly Oscillating Magnetic Field, Apparatus for Producing, 354, 355.
Ratio of Transformation, 311.
Reluctance, Effect of Character and Dimensions of Circuit on, 209, 210.
———, Magnetic, 206.
——— of Human Body, 218, 219, 220.
———, Unit of, 211.
Reluctivity, 208.
Remanent Flux, 202.
Residual Magnetism, 201, 202.
Resistance, Calculation of, 69, 70.
———, Dielectric, 167, 168.
———, Electric, 63 to 79.
———, Electric, Definition of, 63, 64.
———, Electric, of Flesh, 75.
———, Electric, of Human Body, 76, 77, 78.
———, Electrostatic, 156.
———, Joint, 73.

Resistance, Magnetic, 206.
———, Specific, 67.
———, Unit of, Electric, 64.
Resistances, Parallel-Connected, 73.
———, Series-Connected, 72.
Resistivities, Effect of Temperature on, 71.
———, Table of, 68.
Resistivity, Definition of, 67.
———, Effect of Impurity on, 71, 72.
——— of Water, 71.
Rheostat, Carbon, 323, 324, 325.
———, Carbon Pressure, 326.
———, Water, 327, 328.
Rheostats, 321 to 328.
Ribbon Vibrator for Induction Coil, 276, 277.
Rubber of Frictional Electric Machines, 141.
Rule for Direction of Induced E. M. F., 225.

S

Scale, Mirror, Galvanometer, 102, 103.
Secondary Coil, 234.
——— Induced E. M. F. of Medical Induction Coil at High Frequency under Load, 273.
——— of Induction Coil, Inductance of, 264.
——— or Storage Cell, Forms of, 55 to 62.
Self-exciting Dynamos, 302.
Self-Induction, 222, 229, 230, 332.

Self-Induction, Counter Electromotive Force of, 229, 230.
Sensitive Mirror Galvanometer, 102, 103, 104.
Separately-Excited Dynamos, 302.
Series-Connected Resistances, 72.
Series Connection of Voltaic Cells, 50.
Short Circuit, Definition of, 87.
Silent Discharge, 331, 332.
Silver-Chloride Cell, E. M. F. of, 38.
——— Cells, Battery of, 39.
——— Portable Battery, 53.
——— Voltaic Cell, 36, 37, 38.
Single-Fluid Voltaic Cells, 30.
Sinusoidal E. M. F., 121.
——— Wave, 120, 121.
Sources, Electric, Classification of, 26.
Skin Effect, 350.
Snare, Platinum, Cautery, 320, 321.
South Pole of Magnet, 191.
Sparking Distance Through Air-Gap, 140.
Specific Resistance, 67.
Static Breeze, 331.
——— Induced Currents, 344.
Step-Down Transformer, 309.
Storage Cell, Chloride, 56, 57.
——— Cell, Definition of, 54.
——— Cell, Grid of, 56.

Storage Cell, Julien, 59.
——— or Secondary Cells, Forms of, 55 to 62.
Symmetrical E. M. F., 118.
——— Wave of E. M. F., 119.

T

Table of Resistivities, 68.
Temperature, Effect of on Resistivities, 71.
Therapeutic Uses of Electricity, Dangers in, 365 to 372.
Toepler-Holtz Influence Machine, Construction of, 172.
——— Machine, Operation of, 173, 174, 175.
Transformation, Ratio of, 311.
Transformer, Alternating-Current, 309, 310.
———, Step-Down, 309.
Transformers, Motors and Dynamos, 299 to 328.
Tregohm, Definition of, 66.
Two-Part Commutator, Diagram of, 244, 245.

U

Unit of E. M. F., 28.
——— of Electric Activity, 126.
——— of Electric Resistance, 64.
——— of Electric Work, 126.
——— of Magnetic Flux, 211.

Unit of Magnetic Intensity, 213.
——— of Magneto-Motive Force, 207.
——— of Mechanical Activity, 125.
——— of Quantity of Electricity, 80.
——— of Rate of Electric Flow, 81.
——— of Reluctance, 211.
——— of Work, 124.
Universal Ether, 14.

V

Varieties of Electromotive Force, 107 to 123.
——— of Magnetic Circuit, 196.
Vibrator, Adjustable for Faradic Coil, 274.
———, Mechanical, 335, 336.
Volt, Definition of, 28.
Volta, 6.
Volta's Contact Theory, 6, 7.
Voltaic Battery, Definition of, 50.
Voltaic Cell, Bi-Chromate, 41, 42.
——— Cell, Edison-Lalande, 43, 44, 45.
——— Cell, Elements of, 31.
——— Cell, Exciting Liquid of, 30.
——— Cell, Grenet, 41, 42.
——— Cell, Leclanché, 27, 28.
——— Cell, Negative Plate of, 31.
——— Cell, Partz Gravity, 46, 47.
——— Cell, Polarization of, 32.
——— Cell, Positive Plate of, 31.

INDEX. 401

Voltaic Cell, Silver Chloride Form of, 36, 37, 38.
——— Cells, Connection of, in Series, 50.
——— Cells, Double-Fluid, 30.
——— Cells, Non-Polarizable, 32.
——— Cells, Single-Fluid, 30.
——— Cells, Zinc-Carbon, 40, 41.
——— Couple, Definition of, 30.
——— Dry Cell, 48, 49.
——— Elements, 31.
——— Elements, Varieties of, 33.
——— Pair, Definition of, 30.
Volt-Coulomb, Definition of, 126.
Voltmeter, Definition of, 129.
———, Description of, 131.

W

Water-Gramme-Degree-Centigrade, 135.
———, Resistivity of, 71.
——— Rheostat, 327, 328.
Watt, Definition of, 126.
Waves, Sinusoidal, 120, 121.
Weber, Definition of, 211.
Wimshurst Electrical Machine, 182, 183.
Work and Activity, Electric, 124 to 137.
———, Electric, Unit of, 125.
——— Rate of Doing, 125.
———, Unit of Electrical, 126.

Y

Yoke of Electromagnet, 203.

Z

Zinc-Carbon Cell, E. M. F. of, **41**.
―――― Voltaic Cells, 40, 41.

Elementary Electro-Technical Series.

BY

EDWIN J. HOUSTON, Ph.D. and A. E. KENNELLY, D.Sc.

Alternating Electric Currents,
Electric Heating,
Electromagnetism,
Electricity in Electro-Therapeutics,
Electric Arc Lighting,
Electric Incandescent Lighting,
Electric Motors,
Electric Street Railways,
Electric Telephony,
Electric Telegraphy.

Cloth, profusely illustrated.   *Price, $1.00 per volume.*

The above volumes have been prepared to satisfy a demand which exists on the part of the general public for reliable information relating to the various branches of electro-technics. In them will be found concise and authoritative information concerning the several departments of electrical science treated, and the reputation of the authors, and their recognized ability as writers, are a sufficient guarantee as to the accuracy and reliability of the statements. The entire issue, although published in a series of ten volumes, is, nevertheless so prepared that each volume is complete in itself, and can be understood independently of the others. The books are well printed on paper of special quality, profusely illustrated, and handsomely bound in covers of a special design.

THE W. J. JOHNSTON COMPANY, Publishers,
253 BROADWAY, NEW YORK.

THIRD EDITION. GREATLY ENLARGED.

A DICTIONARY OF
Electrical Words, Terms, and Phrases.

By EDWIN J. HOUSTON, Ph.D. (Princeton).

AUTHOR OF
"Advanced Primers of Electricity"; "Electricity One Hundred Years Ago and To-day," etc., etc.

Cloth, 667 large octavo pages, 582 illustrations, Price, $5.00.

Some idea of the scope of this important work and of the immense amount of labor involved in it, may be formed when it is stated that it contains definitions of about 6000 distinct words, terms, or phrases. The dictionary is not a mere word-book; the words, terms, and phrases are invariably followed by a short, concise definition, giving the sense in which they are correctly employed, and a general statement of the principles of electrical science on which the definition is founded. Each of the great classes or divisions of electrical investigation or utilization comes under careful and exhaustive treatment; and while close attention is given to the more settled and hackneyed phraseology of the older branches of work, the newer words and the novel departments they belong to are not less thoroughly handled. Every source of information has been referred to, and while libraries have been ransacked, the notebook of the laboratory and the catalogue of the wareroom have not been forgotten or neglected. So far has the work been carried in respect to the policy of inclusion that the book has been brought down to date by means of an appendix, in which are placed the very newest words, as well as many whose rareness of use had consigned them to obscurity and oblivion.

Copies of this or any other electrical book published will be sent by mail, POSTAGE PREPAID, *to any address in the world, on receipt of price.*

The W. J. Johnston Company, Publishers,
253 BROADWAY, NEW YORK.

AN ILLUSTRATED WEEKLY RECORD OF ELECTRIC RAILWAY PRACTICE AND DEVELOPMENT.

Established January 1, 1886.

THE ONLY ELECTRIC RAILWAY PUBLICATION IN THE WORLD.

As the only publication in the world devoted to the electric railway industry, and the only journal adequately treating the numerous technical features involved in its modern development and practice, the ELECTRIC RAILWAY GAZETTE aims worthily to represent the activity and progressiveness of the important interests to which it is devoted.

Presenting all the news every week, and describing current improvements and developments immediately upon being brought forward, its pages offer to those engaged in the electric railway field the timely advantages enjoyed in other active and important branches of modern industry.

Subscription in advance, One Year, $3.00,
In the United States, Canada or Mexico;
Foreign Countries, $5.00.

The W. J. Johnston Company,
253 BROADWAY, NEW YORK.

THE PIONEER ELECTRICAL JOURNAL OF AMERICA.

Read Wherever the English Language is Spoken.

The Electrical World

is the largest, most handsomely illustrated, and most widely circulated electrical journal in the world.

It should be read not only by every ambitious electrician anxious to rise in his profession, but by every intelligent American.

It is noted for its ability, enterprise, independence and honesty. For thoroughness, candor and progressive spirit it stands in the foremost rank of special journalism.

Always abreast of the times, its treatment of everything relating to the practical and scientific development of electrical knowledge is comprehensive and authoritative. Among its many features is a weekly *Digest of Current Technical Electrical Literature*, which gives a complete *résumé* of current original contributions to electrical literature appearing in other journals the world over.

Subscription { including postage in the U. S., Canada, or Mexico, } **$3 a Year.**

May be ordered of any Newsdealer at 10 cents a week.

Cloth Binders for THE ELECTRICAL WORLD postpaid, $1.00.

The W. J. Johnston Company, Publishers,
253 BROADWAY, NEW YORK.

www.ingramcontent.com/pod-product-compliance
Lightning Source LLC
Chambersburg PA
CBHW050843300426
44111CB00010B/1114